BRITISH COALMINERS
IN THE NINETEENTH CENTURY

A SOCIAL HISTORY

Other books by John Benson

Studies in the Yorkshire Coal Industry, Manchester 1976
   (with R. G. Neville).
Rotherham as it was, Nelson 1976 (with R. G. Neville).
Walsall as it was, Nelson 1978 (with T. J. Raybould).
The West Riding: unique photographs of a bygone age,
Clapham via Lancaster 1978.

JOHN BENSON

BRITISH COALMINERS IN THE NINETEENTH CENTURY:

A SOCIAL HISTORY

HM

HOLMES & MEIER PUBLISHERS, INC.
NEW YORK

First published in the United States of America 1980 by
HOLMES & MEIER PUBLISHERS, INC.
30 Irving Place, New York, N.Y. 10003

Copyright © 1980 John Benson

Library of Congress Cataloging in Publication Data

Benson, John.
British coal-miners in the nineteenth century.

Bibliography: p. 262—272
Includes index.
1. Coal-miners — Great Britain — History.   I. Title.
HD8039,M62G723   941.08   80—11338

ISBN 0—8419—0592—4

Printed and bound in Great Britain

To Clare

Contents

List of Maps

Acknowledgements

I welcome this opportunity to acknowledge the help I have received in the preparation of this book. I am grateful to Dr Kenneth D. Brown, of the Queen's University of Belfast, with whom I first discussed the idea of the book and who subsequently read the whole of the first draft and to Dr W. R. Garside, of the University of Birmingham, with whom I discussed some of the general issues raised by the study. Chapter 1 has benefited from the comments of Professor A. J. Taylor of the University of Leeds, as has Chapter 5 from those of Dr Angela V. John of Thames Polytechnic. I am also pleased to acknowledge the financial support of the Social Science Research Council which enabled me to undertake some of the research necessary for Chapters 4 and 7.

My other debts lie nearer to home. The Polytechnic, Wolverhampton allowed me a term's study leave to complete the book; Mr John Hunter took over some of my teaching commitments; Mr Peter McLeay drew the maps and Mrs Joan Thomas typed the final draft of the manuscript. I have benefited very much from the encouragement — and criticisms — of my colleagues, Dr G. W. Bernard, Dr Eric Taylor and Mr Harvey Woolf, all of whom took the trouble to read and discuss my work with me. My greatest debt, however, is to my wife, Clare, without whose support and encouragement this book would not have been completed.

John Benson
Wolverhampton
December 1978

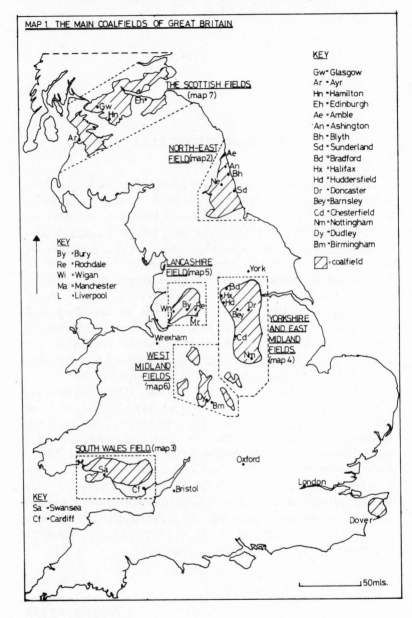

MAP 1. THE MAIN COALFIELDS OF GREAT BRITAIN.

THE SCOTTISH FIELDS (map 7)

NORTH-EAST FIELD (map 2)

LANCASHIRE FIELD (map 5)

YORKSHIRE AND EAST MIDLAND FIELDS (map 4)

WEST MIDLAND FIELDS (map 6)

SOUTH WALES FIELD (map 3)

KEY

Gw = Glasgow
Ar = Ayr
Hn = Hamilton
Eh = Edinburgh
Ae = Amble
An = Ashington
Bh = Blyth
Sd = Sunderland
Bd = Bradford
Hx = Halifax
Hd = Huddersfield
Dr = Doncaster
Bey = Barnsley
Cd = Chesterfield
Nm = Nottingham
Dy = Dudley
Bm = Birmingham

⬚ = coalfield

KEY

By = Bury
Re = Rochdale
Wi = Wigan
Ma = Manchester
L = Liverpool

KEY

Sa = Swansea
Cf = Cardiff

York

Wrexham

Oxford

Bristol

London

Dover

50 mls.

Introduction

Popular misconceptions about the miners have been extremely deep rooted, not least because miners were always seen at their worst. They went to work fresh and clean but this was usually early in the morning, long before most other people were up and about. When they returned home, on the other hand, it was in broad daylight and their pit-dirt stood out in sharp relief. Nor did the introduction of multiple-shift working at the end of the century make them any less conspicuous; for while the new system meant that more people saw the miners as they went to work at various times during the day, 'clean' miners were not necessarily recognisable as such. It must be remembered too that under the new system the miners were also seen more often as they returned from work. Thus in 1910 six miners pleaded guilty under Glasgow Corporation by-laws to travelling on a tram 'in clothing which in the opinion of the conductor might soil or injure the car or dress of the passengers'.[1]

If small groups of miners were viewed with some distaste, large gatherings were regarded with positive dread. When it was heard that the miners were coming, windows were boarded up and people stayed indoors. On at least one occasion, during a dispute over wages in the Shropshire coalfield in 1820, little business could be done at Wellington market because a visit from the colliers was anticipated.[2] Union occasions were particularly unpopular. The *Durham Chronicle* was surprised that the 1872 miners' gala passed off peacefully and that 'the miners ... behaved themselves very well'.[3] When the Lancashire Miners' Federation decided to hold its first gala at Southport in 1890, 'There were a number of residents ... who were afraid and left the town for the week-end.

1

They said that so great a crowd of rough miners would certainly do great damage in the town, and they were afraid that their very nice recreation ground would be destroyed and ruined."[4]

Such fears were not confined to the coalfields. What little was known in the rest of the country about the miners was not reassuring. They worked naked or semi-naked in a dark, dirty, dangerous world hundreds and even thousands of feet below the surface. Indeed it was commonly believed that miners lived underground. As a young man, the Durham union leader, John Wilson, was in a Kent public house when

> the barman wanted to learn who I was, and he put the question. He was told I was a pitman. [Miner is a more modern term.] Pressing for more information he inquired how long I had been down the pit. "Seven years," was the answer. In most surprised tones he said, "Have you not been up until now?" I was surprised at him, and replied, "Yes, every day except on rare occasions." "Why, I thought you pitmen lived down there always," said the querist. It was not long before I gathered from many other quarters that he was not alone in his ideas, for there was a generally held opinion that the coals their ships brought home were dug out of the bowels of the earth by a class of people who were little removed from barbarism, and whose home was down in the eternal darkness.[5]

These misapprehensions were not of course shared by those familiar with the industry. But even coal-mining experts took a jaundiced view of the colliery population.

> The typical coal miner is not always an attractive individual on the outside, but an uncouth and unprepossessing exterior often hides a strong and resolute character and a kindly disposition. Accustomed to face stern and disageeable realities in his daily work, he is real and genuine in his feelings and conduct, and has a robust individuality of his own. Self-reliant and independent, he is no respector of persons, but his respect, once gained, is sincere and lasting...
>
> Strenuous work with liability to danger tends to sharpen a man's faculties and energies, and the miner is usually a

virile and wide-awake sort of person. Of all classes of labour, he is the most grasping and the most combative, the sturdiest fighter in the industrial field, always asking for more...

They show their best side as individuals and their worst when acting together in a corporate capacity, being sometimes misled and misrepresented by their noisiest spokesmen. It is not unusual to see individual excellence coexisting with corporate mediocrity.[6]

The miner thought only of the present: 'It is the first duty of every man to make provision for his family,' argued the *Mining Journal* in 1857, 'and upon no one of the labouring classes is the duty more incumbent than upon miners, for their lives, to use a technical phrase, are doubly hazardous; yet, according to the official returns they would appear to be far behind the rest of the population in providing for themselves and families against accidents.[7]

Modern historians are more circumspect in their language; yet despite the plethora of recent research into labour history, the stereotype of the typical miner has changed scarcely at all since the beginning of the century.[8] The view of the nineteenth-century miner commonly held today is derived essentially from the standard trade union histories, histories which for all their other strengths, are not concerned primarily with the social life of the mass of the miners.[9] Nonetheless they do present a view of the ordinary, rank and file worker which appears to be both consistent and convincing. 'In many places' in the first half of the century 'miners lived a life of barbaric isolation in hovels far worse than those any other section of the community would tolerate.'[10] 'Most of them lived in villages or small towns where almost everyone was engaged in or dependent upon the coal industry, and in consequence mining communities exhibited an exceptional degree of social and occupational homogeneity.'[11] 'There was, no doubt, a streak of Epicureanism in the miner's philosophy, which has not altogether disappeared. Never knowing when bad times, or even fatal accident, would befall him, he tended to enjoy himself whilst he could.... The question of the miner's inability to save is linked with his propensity to 'play'...The basic needs of the miner and his family were easily satisfied in the 'sixties. His pint of beer was comparatively cheap and beyond that his pleasures were few and simple.'[12] The theme is taken up by other historians.

> The brutalized conditions of the miner was reflected in the dirt, neglect and habitual drunkenness prevailing in many mining villages...violence was never far from the miner's life, either at work or leisure. His basic needs remained simple and his tastes were largely customary, hence the increase in absenteeism during booms....Habitual drunkenness led to 'irregular habits' and drained the miner's meagre income, impairing his ability to create and sustain trade unionism.[13]

Such a view of the nineteenth-century miner seems to contradict much of what is known about life in the coalfields and my major aim in writing this book has been to challenge this stereotype of the thriftless and irresponsible miner.[14] Indeed the recurring theme of this study will be the way in which the miners and their wives worked, with little encouragement, to ameliorate the very difficult conditions in which they found themselves. Here then lies my chief justification for adding to the already overwhelming number of works on the coal industry and its labour force. My second objective – and justification – has been to provide a much needed, up-to-date and reasonably comprehensive survey of the social life of the mining population in the nineteenth and early twentieth centuries: the types of homes in which they lived, the ways in which they earned their living, received their pay, spent their money, looked after their wives and children and protected themselves against the vagaries of an uncertain life. In doing this, I have tried to indentify the most serious gaps in our knowledge of life in the coalfields and to indicate those areas and problems that future research might most helpfully seek to elucidate.

The structure of the book is as follows. The first three chapters delineate the economic and social conditions in which the nineteenth and early twentieth century coal-mining community had to live and work. Chapter 1 describes the growth and structure of the coal industry nationally and examines the differing rates of development of the various regional coalfields. It also identifies features like colliery size and concentration of ownership which reappear later as important determinants of social life. In chapter 2 attention switches to the miners' work experience. Emphasis is laid both on differing regional patterns and on the wide range of jobs and skills required in what is often thought to be a monolithic, hewer-dominated industry. Chapter 3 turns to the vexed question of miners' wages. Particular attention is paid to the wide diversity of earnings, to their uncertainty and, so far as is possible, to their movement over the period under examination. The remainder of the book confronts more directly the popular stereotype of the intemperate and improvident miner. The major argument in chapter 4 suggests that the proliferation of small collier-owned shops, the growth of the co-operative movement and the spread of owner-occupation all throw doubt on the easy generalisations that identify miners with lack of foresight. Chapter 5 is probably the least satisfactory in the book. Very little is still known about the ways in which mining families operated; about their diet, courtship, illegitimacy, patterns of child rearing and treatment of the old. What is known, however, does tend to support the thesis that I am propounding. Families were becoming smaller and more settled and it does appear that in this, as in so many other areas of their social life, miners and their wives were making determined and increasing efforts to overcome their difficulties. Chapter 6 examines the miners' reputation for excessive drinking and goes on to maintain that the range of leisure pursuits popular in the coalfields was much wider than is commonly believed. Chapter 7 is in some ways the most important in the book. For it is here, in the analysis of insurance company, friendly society and trade union membership, that it is possible to add statistical support to my thesis that the miners' reputation for irresponsibility and lack of foresight is founded not on fact, but on lack of knowledge and misunderstanding.

1

The Growth of the Industry

Even the most unobservant of contemporary commentators could hardly fail to notice the enormous expansion of the British coal industry during the nineteenth century. At home and abroad there was an unprecedented demand for coal to warm homes, drive railways and steamships and to power factories and workshops. The result was that output multiplied more than twenty-eight-fold from about ten million tons in 1800, to 30 million in 1840, almost 150 million in 1880 and on to a record 287 million tons on the eve of the First World War.[1] This leaping output, the basis of Britain's industrial, economic and commercial power, was made possible by a number of changes within the industry itself. One reason, as will be seen in the next chapter, was to be found in developments in mining technology. Another derived from the changing economic structure of the industry. Until around 1840 few mines were worked by more than about fifty men — not many more than would be found on a good sized farm. But as deeper seams were reached so the pits grew in size. By 1854 the typical British pit was employing about seventy men, a figure which by the end of the period had risen to well over 300.[2] At the same time the industry came to be dominated by large, heavily capitalised firms and, increasingly in the second half of the nineteenth century, by companies working several collieries and employing large numbers of men. As early as the third quarter of the century the Earl of Durham owned eight collieries, that 'Goliath of the Manchester coalfield', Andrew Knowles and Son, eleven, and the giant Wigan Coal and Iron Company twenty-nine.[3] This growing concentration of ownership and production was a significant development in the history of the industry and one

6

industry was severely depressed with the pit-head price of coal — the most reliable indicator of general prosperity — standing at just 5s. 10d. (29 pence) per ton. Four years later the situation had changed dramatically. Output, wages and employment had all increased and pit-head prices peaked at 10s. 10d. (54 pence). Yet by 1905 the pit-head price had fallen to 6s. 11d. (35 pence) and 'coalmining was substantially back where it had been in 1896'.[14] But here again caution is necessary for 'Even in the most profitable period, some collieries made losses; and in the most depressed period some collieries made profits.'[15]

The heterogeneity of coal and its markets, the trade cycle and climatic variation combined with differences in geology, in management techniques and in labour relations make all generalisations well nigh impossible. In fact diversity is perhaps the outstanding characteristic of the nineteenth-century coal industry and mining communities and explains why 'the history of coal mining is to be approached as much at the regional as at the national level'.[16] Each coalfield has its own distinctive history. But for convenience they may be divided into three categories; first, the major coalfields of the North-East, Lancashire and Cheshire, Yorkshire, the Midlands and South Wales; second, the smaller fields of the West Midlands and East and West Scotland; and finally, the minor fields, Cumberland, Kent, the South-West, North Wales and Northern Ireland.

The old established North-East coalfield of Northumberland and Durham was pre-eminent (see Map 2). In 1800 these two counties employed about 13.5 thousand men, a third of all British miners, and produced 2.5 million tons of coal a year, a quarter of the national output. Thereafter output and employment grew rapidly until in 1911 over 200,000 miners were moving 56 million tons a year, most of it — and nearly all the Durham steam and gas coal — for export. From mid-century, the North-East accounted for about a fifth of British coal production and gave employment to a fifth of all miners. Slowly but inexorably, however, this dominance was being challenged by the growth of mining in Yorkshire, the Midlands and South Wales. But it was not until the very end of the period that the North-East was overtaken by South Wales.

In both Northumberland and Durham coal mining was

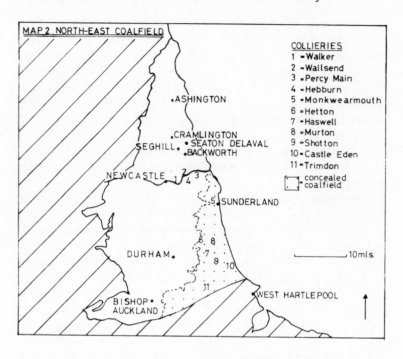

MAP 2 NORTH-EAST COALFIELD

COLLIERIES
1 = Walker
2 = Wallsend
3 = Percy Main
4 = Hebburn
5 = Monkwearmouth
6 = Hetton
7 = Haswell
8 = Murton
9 = Shotton
10 = Castle Eden
11 = Trimdon
= concealed coalfield

.ASHINGTON

.CRAMLINGTON
SEGHILL .SEATON DELAVAL
.BACKWORTH

NEWCASTLE   1 3
           4

5. SUNDERLAND

DURHAM.   6 8
         7
          9
           10

11

BISHOP .
AUCKLAND

WEST HARTLEPOOL

10 mls.

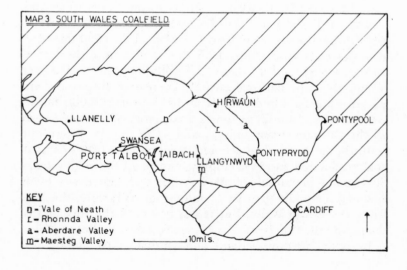

MAP 3 SOUTH WALES COALFIELD.

HIRWAUN

.LLANELLY   n   a   PONTYPOOL
                      r
       SWANSEA
PORT TALBOT .TAIBACH   PONTYPRYDD
              .LLANGYNWYD
                m

CARDIFF

KEY
n = Vale of Neath
r = Rhonnda Valley
a = Aberdare Valley
m = Maesteg Valley

10 mls.

characterised by a high level of investment and, particularly in Durham, by large pits and powerful employers. It was estimated that the average cost of 'winning and sinking' a North-Eastern colliery in 1830 was in the order of £60,000 and twenty years later there was something like £24 million invested in the region's coal industry.[17] Such heavy investment was made by powerful owners to enable them to work large collieries. In the early part of the century North-Eastern pits were by far the largest in the country; at a time when most mines employed fewer than fifty workers there were already a dozen pits in Northumberland and Durham with four times that number. In 1841 there were twenty North-East collieries with more than 200 men, a figure which twenty years later had become the average complement of all pits in South Durham. By 1884 the typical South Durham pit was employing over 300 men and the typical Durham owner was running more than two collieries. By 1910 he was running three, with Pease and Partners owning nine, the Joicey Company twelve and the Lambton Company fifteen.[18] So the North-East pitman, especially in Durham, was more and more likely to find himself working in a large pit, just one of a number belonging to his employer.

The enormous expansion in coal production during the nineteenth century inevitably exhausted the more easily worked seams and led everywhere to the more intensive working of existing fields and/or the development of new, more difficult and more dangerous concealed coalfields. This development may be seen clearly in the North-East. The outcrops and shallow seams round Newcastle had been worked for so long that well before the end of the eighteenth century the industry began to move east along the Tyne Valley, opening house sale pits at, for example, Walker (1762), Wallsend (1778), Percy Main (1790) and Hebburn (1792-4). But still 'In 1800 Northumberland and Durham were rural counties into which the coal industry intruded along the rivers Tyne and Wear and to some extent along the coast.'[19] By the 1820s the Tyneside collieries were in decline. So in Northumberland new pits were opened to work the steam coals at Beckworth and Seghill, then at Cramlington and Seaton Delaval and finally, at the turn of the century, round Ashington and up to Amble.[20] Even greater was the expansion south of the Tyne. Collieries were sunk to the house coals that lay beneath the magnesian limestone of the East Durham plateau. Monkwearmouth began to raise coal in 1831, to be followed by Hetton (1832), Haswell (1835), Murton (1838), Shotton (1841), Castle Eden (1842) and Trimdon (1843).[21] Then with the coming of the railways there began the exploitation of the coking coals of south-west (or mid) Durham, between Durham City and Bishop Auckland, the growth point after 1850 of the entire North-East coalfield.[22] By this time the whole economic, social and industrial structure of the North-East was founded on coal. In the words of the Geordie pit-man's song:

I tell the truth you may depend:
In Durham or Northumberland
No trade in them could ever stand
If it were not for the coal trade.[23]

The situation in Lancashire and Cheshire was very different (see Map 5). Here the economy was more broadly based and small-scale mining lingered longer than in any other coalfield with the exception of South Staffordshire. Small pits were specially common around Rochdale and Bury in the north-east of the county where it was possible to get coal from shallow workings or from drifts (tunnels) dug into the sides of the Pennines.[24] Thus in the early 1860s, when the typical South Durham pit was employing over 200 men, its north-east Lancashire counterpart was worked by well under half that number and even at the beginning of this century there remained sixteen north-east Lancashire collieries (like the tiny three-man Bamford Closes colliery at Wolstenhome) employing fewer than twenty workers.[25]

The picture did not really begin to change until the middle of the nineteenth century when the industry started to move from these northern outcrops to the more deeply concealed seams in the south-west of the county. Thereafter Lancashire collieries varied enormously in size and productive capacity. As the district's inspector of mines, Joseph Dickinson, commented in 1854, 'The smaller ones employ only a few hands, and the larger, from one thousand up to fifteen hundred persons. The output of coal per colliery varies from a few tons to a third of a million tons per annum.'[26] In fact Lancashire saw the deepest sinkings and the growth of some of the largest firms in the country. By the 1860s the Astley New Pit at Dukinfield had reached 683 yards and the county housed such giants of the industry as the Bridgewater Trustees and the Richard Evans, George Hargreaves and Wigan Coal and Iron Companies so that, for example, five major companies were able to control over half the output of the whole of the Wigan coalfield.[27] The labour force at each colliery was also getting larger. In 1854 the typical Lancashire colliery employed 85 people, a fifth above the national average. By the beginning of the present century the disparity was even more marked with the average Lancashire colliery labour force of 423 standing more than two-fifths above the average for the whole of the United Kingdom.

The Lancashire and Cheshire coal industry grew almost twenty-fold during the nineteenth century. But it was constantly losing ground relative to the other major coalfields. In 1860 it contributed over a seventh of national output and was the third largest producer behind the West Midlands and the North-East. In 1908 it had less than a tenth of total production and lay only sixth among the coalfields, having been further overtaken by Scotland, the Midlands and South Wales.

Across the Pennines in Yorkshire there was a movement away from the outcrop similar to that which had taken place in Lancashire and the North-East (see Map 4). In the early nineteenth century Yorkshire coal-mining was confined largely to the west of the county around Bradford, Halifax and Huddersfield where small, often family-owned pits worked shallow seams for the local textile industry, the first six-inch edition of the Ordnance Survey published in 1854 showing more than 200 pits in the Aire and Calder valleys alone.[28] But from the 1830s the working out of these collieries, the construction of a basic railway network and the adoption of new mining techniques led to the development of the South Yorkshire coalfield. At first the industry concentrated on two major seams, the Silkstone and the Barnsley Bed, where these lay fairly near the surface round Barnsley. The first large collieries to exploit the Barnsley Bed were Mount

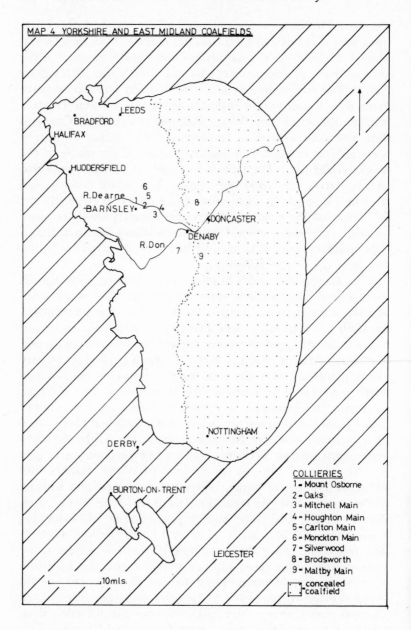

MAP 4 YORKSHIRE AND EAST MIDLAND COALFIELDS.

LEEDS

BRADFORD

HALIFAX

HUDDERSFIELD

R.Dearne

BARNSLEY

R.Don

DONCASTER

DENABY

NOTTINGHAM

DERBY

BURTON-ON-TRENT

LEICESTER

10mls.

COLLIERIES.
1 = Mount Osborne
2 = Oaks
3 = Mitchell Main
4 = Houghton Main
5 = Carlton Main
6 = Monckton Main
7 = Silverwood
8 = Brodsworth
9 = Maltby Main
concealed coalfield

Osborne (1838) and the Oaks (1838-40), the latter at the unprecedented depth (for Yorkshire) of 288 yards. Thirty years later the Barnsley Bed was being worked at over 420 yards at Denaby, seven miles from Doncaster and just west of the river Dearne, but it was not until the boom of the early 1870s that it proved profitable to work the Barnsley Bed as it plunged deeper and deeper east of the Dearne: then the Mitchell Main colliery was sunk to 307 yards in 1871, followed closely by Houghton Main at 514 yards in 1873, Carlton Main at 287 yards in 1873-4 and Monckton Main at 475 yards in 1875. It was only at the end of the century that the industry turned finally to the totally concealed coalfield east of Doncaster. Silverwood colliery was opened in 1903, Brodsworth in 1908 and in 1909 Maltby Main was sunk to the Barnsley Bed at 820 yards.[29]

These South Yorkshire pits sold most of their coal outside the county. In 1870 they were selling domestic coal to London, steam coal to Grimsby (for export), engine coal to Manchester and the East Manchester connurbation and coking coal to Lincolnshire and other iron-smelting districts.[30] South Yorkshire pits were also larger, more heavily capitalised and more profitable than the older collieries in the west of the county. By 1860-1 the average South Yorkshire colliery was employing eighty-three workers, twice the number to be found in the west. The cost of opening these large new pits was correspondingly high, each of the eight collieries sunk to the Silkstone and Barnsley Bed seams round Barnsley in the 1880s costing between £100,000 and £150,000.[31] The concentration of ownership too was far greater in the south. Chapeltown had the Newton Chambers Company, Wath-upon-Dearne the Manvers Main Colliery Company and Nether Hoyland the Hoyland and Silkstone Coal and Coke Company. The tendency, already obvious by the middle of the century, was for South Yorkshire pits to be large, run by joint stock partnerships and difficult to manage.[32]

The growing dominance of South Yorkshire over the West is reflected in the production figures of the two districts. At mid-century South Yorkshire was producing a third, by 1870 just over 40 per cent and by early this century more than 70 per cent of the county's total output.[33] It was because of this

South Yorkshire expansion that the industry was able to grow so rapidly, both absolutely and in relation to other coalfields. Between 1800 and 1908 employment probably increased more than forty times until a seventh of all British miners were working in the county. Output too multiplied rapidly so that in 1913 Yorkshire was the third largest coal producing region, ahead of Scotland, Lancashire and the Midlands and behind only the North-East and South Wales.

The East Midland district of Derbyshire, Nottinghamshire, Leicestershire and Warwickshire was something of a latecomer to the ranks of the leading British coalfields. For the first sixty years of the last century it never rose above sixth position. But by 1900 it lay fourth, with only Yorkshire, the North-East and South Wales employing more miners or producing more coal. As in all districts, early nineteenth-century mining in the Midlands was to be found in those areas where the coal was most readily accessible. In Warwickshire the industry was confined to the areas between Tamworth and Atherstone and in Leicestershire to the exposed measures round Swannington and Coleorton. There was far more mining in Derbyshire and Nottinghamshire but still the pits clung to the outcrops which ran to the west of a line drawn roughly from Nottingham to Chesterfield. But just as in Yorkshire, growing demand, new mining technology and the coming of the railways made it both possible and profitable to work deeper seams and compete in the rich London market. The sinking of Whitwick colliery in 1824 marked the opening of the concealed field in south Leicestershire. Far more important, there began the development of the concealed Notthinghamshire and Derbyshire coalfield which lay to the east of the previously worked outcrop. A series of large, new pits were sunk around Hucknall in the 1860s and 1870s: Hucknall 1 and 2 (1861-2), Annesley (1865), Bulwell (1867), Bestwood (1871-2) and Newstead (1875). But it was the final decade of the century which saw the really intensive development of the famous Top Hard steam coal (a continuation of the South Yorkshire Barnsley Bed). Among the large collieries opened on the Nottinghamshire-Derbyshire border round Mansfield and to the east of Chesterfield were Bolsover (1890), Warsop Main (1895) and Creswell, Markham and Shirebrook (all in 1896). By 1913, half of all Nottingham-shire coal and 'fully 25 per cent of the Derbyshire production was top-hard'.[34]

Despite this, the labour force of most Midland collieries remained fairly small, hovering about the 150 mark through-out the 1860s, 1870s and 1880s. Eventually, however, the opening of the larger and deeper Top Hard pits and the growth of the industry in Warwickshire did have its effect. Between 1889 and 1894 the labour force of the typical Midland colliery leapt from 151 to 257. By 1908 it was getting on for a thousand. Large numbers of small pits were giving way to a smaller number of relatively large pits. Surprisingly, perhaps, this growing concentration of production was not matched by any real increase in the concentration of ownership. There were of course some large firms, especially in Derbyshire. In 1863 the Staveley Company was employing over 3,000 men and producing a million tons of coal a year, and by the turn of the century the Butterley Company was able to produce nearly three million tons a year from its thirteen pits. But it must be remembered that even at the end of the period the average East Midland colliery owner did not work more than about 1.5 pits.[35]

The latest — and fastest — developer of all the major British coalfields was South Wales (see Map 3). Its coal industry was insignificant at the beginning of the period when it gave em-ployment to about 1,500 people. But it expanded enor-mously until by 1913 it employed more than 233,000 men, a staggering one hundred and fifty-five-fold increase. In 1800 South Wales stood ninth of the twelve coal producing districts; in 1840 it was fourth and by 1880 had risen to second place. Finally, at the very end of the period, it became the premier coalfield. In 1913 a fifth of all miners and a fifth of all coal came from South Wales.

In the early nineteenth century coal was mined only on the south-western and north-eastern rims of the coalfield. In the south-west, in Pembrokeshire, round Llanelly and Swansea and up the Neath Valley, anthracite was produced for domestic consumption and some bituminous coal was mined for the local copper industry. Near Port Talbot the English Copper Company had levels (or tunnels) to supply fuel to their copper smelting works at Taibach. Meanwhile the ironmasters held sway in the north-east of the coalfield. From Hirwaun in East Glamorgan to Pontypool in Monmouthshire they worked the bituminous coal needed to fire their furnaces.[36] At mid-century the most important sector of the coalfield was already that to be found in Monmouthshire and East Glamorgan. Thereafter its dominance became even more marked. The anthracite found to the west of the Vale of Neath finally proved unsuitable for either smelting iron or raising steam. At the same time entrepreneurs were coming to realise the full potential of the high quality steam coals in the middle of the coalfield. Their exploitation was encouraged by the increased demand coming from the introduction of steam-driven ships and by the improvement of local transport facilities – the Taff Vale Railway and the West Bute Dock at Cardiff were both opened in the early 1840s. The industry moved into the Aberdare Valley and then from the 1860s into the deeper seams of the Rhondda.[37] 'In the two decades before 1875 the growth in production from the two valleys accounted for 37 per cent of the total increase in output from the coalfield, while their share in the increase of coal shipments, coastal and foreign, from South Wales was probably over 60 per cent in the period between 1840 and 1854 and about 70 per cent in the following two decades.' From the final quarter of the century the Rhondda took over from the Aberdare Valley as the leading coal producing area in South Wales.[38]

The growing importance of the steam coal trade led to an increase in the size of both ownership and production and thus to changes in the mining communities themselves. At the start of the century there were many very small collieries like the pit at Llangynwyd in the Maesteg Valley which 'in 1790 employed but a single collier and a windsman'. In the early 1840s the majority of mines were still worked from levels rather than from vertical shafts and the average owner still possessed just one colliery at which he employed less than a hundred men and produced well under 50,000 tons a year.[39] By the mid-1870s 'The small collieries still predominated so far as numbers were concerned but they had ceased – in terms of contribution to the total output – to be the characteristic unit of the industry.' Something like three-quarters of South Wales production now came from collieries producing more than 50,000 tons a year. 'The exception of 1840 had come to be rather less than the average of 1875.'[40] By the end of the period the labour force at a typical South Wales mine stood at 376, some 10 per cent above the national average.

The large amounts of capital needed to work the deep steam coal seams of the Aberdare and Rhondda were beyond the reach of most individuals and small partnerships. So within twenty years of the passing of the Limited Liability Act in 1855, nearly half of South Wales production was controlled by limited liability companies, some of which like the Glamorgan Coal Company and the Powell Duffryn Steam Coal Company were destined to become almost household names. By 1916 the Cambrian Combine controlled two-thirds of the entire output of the Rhondda Valley.[41] All over the coalfield it became increasingly common for an employer to own more than one colliery. In the early 1840s, the average South Wales coal owner had been working one colliery; in the 1870s, he was working 1.6. By 1908, he (or more probably it) was working two.

We have considered the five major British coalfields, the North-East, Lancashire and Cheshire, Yorkshire, the Midlands and South Wales, in some detail because they played a dominant and growing role in the nineteenth-century coal industry. In the early nineteenth century, six out of every ten miners — and by its end, eight out of every ten miners — worked in one of these five districts. In contrast, the three middle-ranking coalfields, the East of Scotland, the West of Scotland and the West Midlands, never managed to play such a large part. Indeed their importance tended to decline. At the beginning of the nineteenth century a quarter of British miners were to be found in Scotland or in the West Midlands, but by the end of the period, the number had dropped to a fifth.

For the first seventy to eighty years of the nineteenth century, the largest middle-ranking coalfield was the West Midland district of Shropshire, Worcestershire and Staffordshire (see Map 6). Its output increased from about 1.2 million tons in 1800 to 15 million in 1880, a growth due almost entirely to more intensive working in the north-east of the Black Country at Bilston, Tipton, Coseley, Sedgley and Dudley and later round West Bromwich and Rowley Regis.[42] In this part of South Staffordshire and East Worcestershire the famous Thick (or Ten Yard) Coal was less than sixty yards from the surface; it was easy to work and provided fuel for every domestic and industrial need and especially for the local ironworks.

The Black County coal industry collapsed in the 1870s. The best seams were nearing exhaustion; there was serious flooding round Tipton ('the great flooded district of South Staffordshire'); the iron industry was in decline and there was increasingly severe competition from Cannock Chase and elsewhere. Pits closed, employment dropped and production plummetted. From nine million tons in 1877 output fell to six million in 1886 and three million in 1913.[43] That the collapse of South Staffordshire did not bring the whole West Midland coal industry down with it was due to the growth of North Staffordshire and Cannock Chase during the second half of the century. As soon as the technical difficulties of working the wet, faulted, inclined seams of North Staffordshire were overcome, its share of county production rose rapidly from a fifth in 1856 to 70 per cent at the end of the century.[44] When the Cannock Chase industry moved from the outcrop to the exposed field, it too came to supply an extensive market. 'From a very small proportion of the combined production of the Cannock Chase and South Staffordshire coal-fields in the fifties, the Cannock Chase field was responsible for more than one-third of production in 1880', for about half in 1898 and for some 70 per cent by the outbreak of World War I.[45]

Despite the growth of coal production in Cannock Chase and North Staffordshire, the nineteenth century still saw the serious decline of the West Midlands in relation to other districts. For the first forty years it was the second largest coalfield in the country employing around 15 per cent of the country's coalminers and producing the same proportion of its coal. By the end of the century it had slipped to eighth position with just seven per cent of coal production and mining employment.

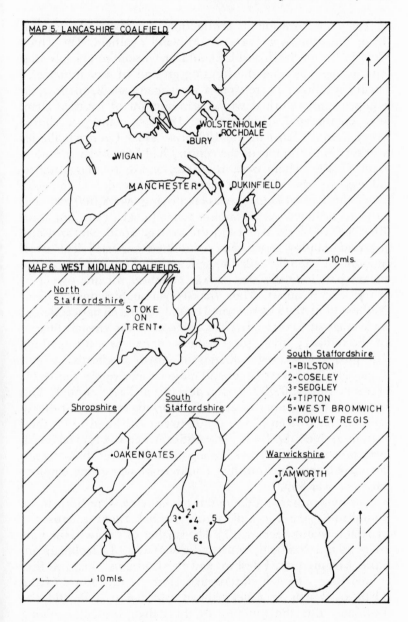

MAP 5. LANCASHIRE COALFIELD

WOLSTENHOLME
ROCHDALE
BURY
WIGAN
MANCHESTER
DUKINFIELD

10 mls.

MAP 6. WEST MIDLAND COALFIELDS.

North
Staffordshire
STOKE
ON
TRENT

Shropshire

South
Staffordshire

South Staffordshire
1 = BILSTON
2 = COSELEY
3 = SEDGLEY
4 = TIPTON
5 = WEST BROMWICH
6 = ROWLEY REGIS

OAKENGATES

Warwickshire
TAMWORTH

1
2
3
4
5
6

10 mls.

The concentration of production and ownership was far from uniform over the West Midland coalfield. The Thick Coal in South Staffordshire could be worked so cheaply that mining here became the hunting ground of any number of small capitalists. There were a great many small, shallow, ill-equipped pits worked with the minimum of capital. There was, commented an observer in 1819, 'a colliery in almost every field'.[46] Most of the 400 or so Black County collieries working in the middle of the century had been opened for less than £3,000. There were, it is true, one or two large undertakings. By the 1870s the unquestioned leader of the coalfield, the Earl of Dudley, was employing over 8,000 men and boys to produce over a million tons of coal a year. But at no time before World War I did the average South Staffordshire pit ever employ more than a hundred miners.[47] The small amount of capital needed to sink a pit in South Staffordshire meant paradoxically that it was common for one owner to work several mines. Thus the sixty-nine pits open around Netherton in 1843 were owned by only nineteen undertakings, an average of 3.6 each. Even in 1889, after the virtual collapse of the industry, the typical colliery concern was still working 1.5 pits.[48] The new collieries of North Staffordshire and Cannock Chase were larger and more expensive. By the 1870s it was taking £150,000 to open new pits at East Cannock. And although the number of collieries working in North Staffordshire declined from 123 to 109 between 1854 and 1870, their output trebled. So by the late 1880s, when the typical South Staffordshire colliery was finding work for fewer than seventy men, its North Staffordshire counterpart was giving employment to over 300.[49]

The two other middle-ranking coalfields were both to be found in Scotland in a discontinuous belt running across the central lowlands (see Map 7). The smaller of the two, the East of Scotland coalfield, consisted essentially of Clackmannanshire, Stirlingshire, Fifeshire and the Lothians. Coal was first mined here where it outcropped close to the shores of the Firth of Forth in Clackmannan, south-east Fife and the Lothians. But the building of the railways, greater mining expertise and the growth in demand (particularly for export) later encouraged large-scale inland mining. Pits were opened at for example Bathgate and Newbattle in the Lothians, near Alloa in Clackmannan, and round Cowdenbeath, Lochgelly, Dunfermline and Kirkaldy in Fife. In the 1870s and 1880s, East Scotland bituminous coal was being exported to Scandinavia, Germany, the Mediterranean and the Baltic. Indeed by the end of the century Fife was the fastest growing coalfield in the whole of the United Kingdom.[50]

Overall, the East Scotland coal industry grew steadily if unspectacularly, multiplying fourteen times (just half the national average) during the century. It never figured among the top six coal producing districts of the United Kingdom and was constantly having to give way to the larger West Scotland field. As in so many other districts the development of new fields was accompanied by an increase in the concentration of both ownership and production and, as will be seen later, by changes in social behaviour. Between 1875 and 1885, the number of concerns raising under 40,000 tons a year dropped from 144 to eighty while the number producing over a quarter of a million tons more than doubled. In the late 1880s, most East Scotland miners worked with nearly 200 other workers and by the present century, the average work force was getting on for 340.[51] But although the landed estate began to withdraw from active involvement in the industry and it was common for firms to describe themselves as companies, the corporate form of ownership was in fact slow to emerge. As late as 1886 only about thirty-five of the 300 concerns in the whole of Scotland were true registered companies. And it was not until the 1890s that East Scotland giants like the Alloa, the Lothian and the Fife Coal Companies really began to dominate the regional coal industry. In 1911 the Fife Coal Company had a labour force of over 10,000 and an annual output of more than two million tons.[52]

The bulk of Scottish output came from the west of the country, from Dunbartonshire, Renfrewshire, Ayrshire and from Lanarkshire, the largest coalfield north of the border. In the early nineteenth century high transport costs restricted mining largely to the immediate vicinity of Glasgow, the major market, and to the Kilmarnock and Mauchline basins in Ayrshire, which were within five miles or so of navigable

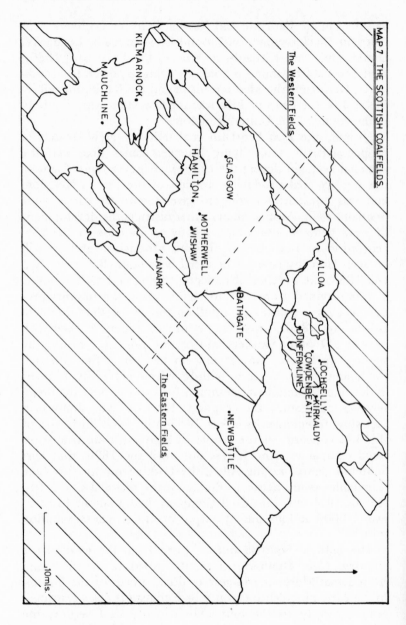

water.[53] By mid-century, transport costs had been slashed by the railways, rich blackband ironstone had been discovered in Lanarkshire, and the Clyde Valley had grown into a major manufacturing and shipping centre.[54] The result was dramatic. In Ayrshire there was intensive development of the Kilmarnock basin and other inland areas. In Lanarkshire pits were opened along the Clyde valley, past Hamilton, Motherwell and Wishaw, almost into Lanark itself.[55] The great boom of the early 1870s saw scores of deep new collieries sunk to the rich seven foot Ell coal between Glasgow and Hamilton. From 1871 to 1877, for instance, fourteen collieries (with thirty-two pits) were opened within a two-mile radius of Hamilton West railway station.[56]

The approximate twenty-eight-fold increase in West of Scotland output during the nineteenth century matched very closely the growth of national production. By 1913 the West of Scotland was the sixth largest coalfield in Great Britain, producing very nearly as much coal as did Lancashire and Cheshire. Despite this impressive growth, the mines themselves were smaller than those in East Scotland. In the late 1880s most West Scotland miners were still working in pits of under 150 men and even by the end of the period the normal labour force barely touched 250.[57] But while the typical owner had only one or two collieries, a few large concerns did come to dominate. By 1870 the Shotts Iron Company and William Dixon each ran ten collieries and both William Baird and Merry and Cunninghame controlled twenty-two. By the beginning of this century William Baird and Company owned over a third of the seventy-five collieries in Ayrshire and, together with the United Collieries of Glasgow, owned twenty-two of the 205 collieries in Lanarkshire.[58]

The minor coalfields need not detain us long for at no time did Cumberland, Kent, the South-West or North Wales and Northern Ireland together employ more than a tenth of British miners. Indeed their importance was on the wane so that by 1880 only six per cent and by 1908 only four per cent of British miners were at work in them. The Kent industry was minute during this period and even by the end of World War I only two collieries were in production.[59] Northern Ireland was no more important; its total output was scarcely more than that of a single large colliery company in England or South Wales.[60] The only minor fields of any consequence were Cumberland, the South-West and North Wales, all of which grew steadily during the course of the nineteenth century. Although expansion of the North Wales coalfield of Denbighshire and Flintshire was restricted by limited local demand, the labour force did rise from about 1,200 at the start of the century to 15,000 on the eve of World War I, when output stood at 3.5 million tons. It remained a small-scale industry; in 1913 there were only fifteen mines whose underground labour force exceeded a hundred.[61] The development of the old-established South-West coalfield of Gloucestershire and Somerset proceeded in similar fashion. The labour force grew from just under 2,000 at the beginning of the period to some 15,000 at its end. It was hard to expand production very greatly because the area was mainly rural, because (especially in the Forest of Dean) the industry was conducted on a small-scale family basis and because the district's house coal faced increasingly severe competition from the booming Midland coalfield.[62] The Cumberland coalfield, with its large employers, extensive pits and deep workings was in many respects, 'the North East in microcosm'. As early as 1813 the Lowthers were employing 600 workers and by 1842 there were more than 1,300 at work in the Lonsdale mines. Not surprisingly Cumberland has been dubbed 'the kingdom where Lord Lonsdale was absolute monarch'. But the irregular and faulted nature of the seams, the distance of many undersea faces from the pit bottom, and the amount of dirt brought up with the coal made Cumberland's bituminous coal expensive to work. These difficulties were aggravated by the county's geographical position which meant that it had to compete in the Irish market with Lancashire house coal as well as with manufacturing coal from Scotland. So the Cumberland coal industry always remained small, never employing more than about 10,000 men or producing much more than two million tons of coal a year.[63]

It is difficult to overemphasise the importance of coalmining in nineteenth and early twentieth-century Britain. With output growing so rapidly, employment rose from about 42,000 in 1800 to over 1,127,000 in 1911 when 'miners were at for example Bathgate and Newbattle in the Lothians, near Alloa in Clackmannan, and round Cowdenbeath, Lochgelly, Dunfermline and Kirkaldy in Fife. In the 1870s and 1880s, East Scotland bituminous coal was being exported to Scandinavia, Germany, the Mediterranean and the Baltic. Indeed by the end of the century Fife was the fastest growing coalfield in the whole of the United Kingdom.[50]

Overall, the East Scotland coal industry grew steadily if unspectacularly, multiplying fourteen times (just half the national average) during the century. It never figured among the top six coal producing districts of the United Kingdom and was constantly having to give way to the larger West Scotland field. As in so many other districts the development of new fields was accompanied by an increase in the concentration of both ownership and production and, as will be seen later, by changes in social behaviour. Between 1875 and 1885, the number of concerns raising under 40,000 tons a year dropped from 144 to eighty while the number producing over a quarter of a million tons more than doubled. In the late 1880s, most East Scotland miners worked with nearly 200 other workers and by the present century, the average work force was getting on for 340.[51] But although the landed estate began to withdraw from active involvement in the industry and it was common for firms to describe themselves as companies, the corporate form of ownership was in fact slow to emerge. As late as 1886 only about thirty-five of the 300 concerns in the whole of Scotland were true registered companies. And it was not until the 1890s that East Scotland giants like the Alloa, the Lothian and the Fife Coal Companies really began to dominate the regional coal industry. In 1911 the Fife Coal Company had a labour force of over 10,000 and an annual output of more than two million tons.[52]

The bulk of Scottish output came from the west of the country, from Dunbartonshire, Renfrewshire, Ayrshire and from Lanarkshire, the largest coalfield north of the border. In the early nineteenth century high transport costs restricted mining largely to the immediate vicinity of Glasgow, the major market, and to the Kilmarnock and Mauchline basins in Ayrshire, which were within five miles or so of navigable

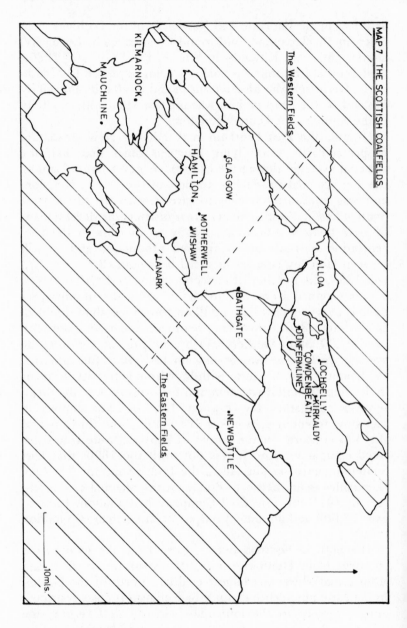

MAP 7   THE SCOTTISH COALFIELDS

almost as numerous as building and construction workers or those engaged in agriculture; they were more numerous than the main branches of transport workers and out-numbered male textile operatives by almost two to one.'[64] Throughout the whole of the second half of the nineteenth century more than one person in twenty looked directly to the coal industry for his livelihood. Well might George Stephenson urge the Lord Chancellor to get up off his woolsack and sit instead on a sack of coals![65]

2

The Miner at Work

Whenever and wherever it was carried out, mining work was far more heterogeneous than is generally supposed. Surface workers, for example, had very low status in mining communities. Indeed to judge from the contempt of their underground colleagues — and the neglect of almost everybody else — they were scarcely regarded as miners at all. Yet the boys and girls, women, craftsmen and 'knocked-up' ex-hewers employed on the surface made up a large, and growing, minority of the total coal-mining labour force. At the end of the period a fifth of all miners, a total of well over 200,000 men, women and children, were working at the pithead.[1]

There are two main reasons for this neglect of surface workers. Perhaps the most important is that their jobs were far less homogeneous than those done underground. Thus while there were just four categories of workers employed underground at the Govan collieries in 1834, above ground there were eight: labourers, blacksmiths, hammermen, iron basket makers, winding-enginemen, wheelrights, masons and banksmen (who supervised the unloading of the tubs).[2] It is not easy to find out much about the labourers who tidied up, stacked the pit props, counted the tokens (recording output), weighed the tubs, moved the wagons, stoked the boilers, worked the tippers and picked, washed and screened (separated) the coal. Nor is it easy to discover a great deal about the handful of winding-enginemen, the two or three banksmen in charge of the cage or even about the whole batch of different tradesmen needed to keep a fair sized pit and its equipment in efficient running order.

It is known however that twice a day the engineman had the lives of all the underground workers in his hands. Early in the century even this highly responsible job was given to children – often with calamitous results. In 1813 one miner died and two were injured when the boy in charge of the engine at Price's Field colliery in Wolverhampton failed to stop the machine in time.[3] But eventually children were excluded from the engine shed and winding became the province of older and more careful men, the one exception to the inferior status generally afforded to surface workers. 'The working aristocrat,' remembers one Yorkshire miner, 'was the winding engine man, in his billycock hat, suit, collar and tie. The winding engine men were always needed at the pit, so they never lost money through being put on short time, and they worked every weekend, too. They even had their own place to wash, up in the winding house.'[4] We know too that for more than half the nineteenth century colliery tradesmen spent nearly all their time on the surface, repairing wagons, boilers and engines, sharpening picks and other tools, seeing to the horses and their tack and maintaining the colliery buildings. But changes in mining technique meant that from about 1860 onwards an increasing number of tradesmen had to go below. With the introduction of the longwall system of working (with a number of men working along a single coal face), masons and bricklayers were needed to build stoppings to re-route the flow of air. And as fixed engines were installed to improve the movement of coal underground, so fitters (and later electricians) had to go down to repair them.[5]

The age, sex and skill of surface workers varied enormously then; from young girls picking dirt from the coal to skilled tradesmen shoeing horses, making wagons and mending engines. This heterogeneity is most important. It goes far towards explaining both the difficulty of unionising surface workers and their consequent neglect by almost all historians of the coal industry and miners' trade unionism.

The second reason for the lack of attention paid to surface work is that it does not have the fascination of underground labour. Working at the pithead was undramatic. But it was heavier, dirtier, more unpleasant and more dangerous than many other jobs above ground. The hours were long, between nine and twelve a day, five or six days a week. Night shifts were common and Sunday work widespread among blacksmiths, joiners and fitters. 'You may hear their tools going clink and clank all the day, regardless of the chimes of the many church bells which are calling us all to the house of God.'[6] It was unhealthy: coal dust got everywhere making everybody 'as black as those who are down in the mines'.[7] At least one South Wales 'man was suffocated . . . by the fumes from the tip at Dowlais'.[8] Because of the expense of building shelters, much of the work had to be done in the open. Even at many large new pits opened at the beginning of this century 'the cabins for the outside workers were simply broken trams pushed up on their ends'.[9]

Working on the surface was only about a fifth as dangerous as it was underground. But even this represented a substantial risk. There were innumerable ruptures and strains. Children were specially at risk; in 1858 'an interesting little girl' hired to pick coal at Tipton colliery in the Black Country was burnt to death when she got too close to some live embers.[10] Before the installation of self-acting fences (fences which closed automatically when the cage descended) many workers of all ages simply fell down the shaft. Hannah Rees for instance was 'dashed to pieces' at the bottom of a deep pit belonging to the Tredegar Iron Company.[11] Getting too near the shaft remained a potent fear: 'The Onsetter Shouted for me', recalls a Durham boy, 'A Big Strong Feller, got me in his Arms and He Held me near Shaft, He Frightened my life out . . .'[12] The heavy coal wagons were always dangerous as well. At Major Ward's Leeds colliery a blacksmith was killed in 1863 when a number of empty wagons ran into the smithy. Twenty years later Jane Eastham of the Hindley Green colliery, St Helens, was passing between the buffers of some stationary wagons when some moving wagons bumped into them and killed her. Indeed between 1852 and 1890 thirty-eight women were killed on the surface by wagons running out of control.[13] However the relatively low liability to injury of surface workers has undoubtedly been one factor determining the lack of interest shown in this part of colliery work. But surface work was far more hazardous than most people imagine. Year upon year the death rate of surface workers exceeded that resulting from the massively publicised underground explosions.[14]

The only feature of surface work to receive any attention at all has been the employment of women and young girls, some of whom became well-known local characters. Sal Madge, for example, started work at Whitehaven at the age of eight and went on to spend sixty years on the pit stop until her death in 1899. But for all the controversy which female employment aroused, the numbers at work were both small and constantly declining. It was always a highly localised affair, being confined to the small, backward pits of the East of Scotland, Shropshire, Cumberland, South Wales and West Lancashire. By the 1880s they numbered in all no more than about 4,000, fewer than five per cent of the total surface labour force.

There were, it is true, periodic local revivals in female employment, at least in the short-term. The exclusion of women from underground work in 1842 led to more women being taken on above ground. In East Scotland, twenty-four of the 117 women working below ground for the Clackmannan Company were moved to the pit bank. Then in the early 1860s the cotton famine stimulated surface employment round Wigan and possibly also in Clackmannan, Fife and Lanarkshire. Finally, at the beginning of this century, some employers in East Scotland and Lancashire turned again to women and girls in the hope that they would prove more reliable and hard-working than male surface workers. Accordingly firms like Thomas Fletcher and Son replaced the men on the picking belts at their Outwood colliery near Manchester with thirty to forty young women.[15]

But these were exceptions. The overall trend in female employment was firmly downwards and, anyway, women and girls had never been employed at all in most districts. Yet concentration on this group has diverted attention from the great majority of the surface labour force. Some little is known about five per cent of pit top workers, the women and girls, but next to nothing about the other 95 per cent, the men and boys. The lack of any serious research on the history of surface work constitutes a serious gap in our knowledge of the nineteenth-century coal industry. It is rather hackneyed to make a plea for further research. But in this case the plea is surely justified; we need to know more, a great deal more, about that almost totally forgotten miner, the nineteenth-century surface worker.

We know of course far more about the lot of the underground worker. Underground work was different from all other. The pit was a world apart — dark, dusty, insanitary and unpredictable — and very often hot, stuffy, wet and cramped as well. Even children brought up in mining communities and used to hearing 'other men and boys describing their work in every-day talk', found the transition to pit work traumatic. The young son of one Black Country collier was scared to work below because 'they kill people down there'. When the small Durham boy, George Parkinson, followed his father underground in 1837 he found 'Everything was new and strange' and he had to 'be initiated into . . . the mysteries and the manly phraseology of a pit-boy's life'.[16] This could be a daunting process. A Durham pitman remembers that during his first shift, the putters (who moved the wagons) 'exceeding horse play fashion, forced me down and held me while another poured oil from midgies (lamps) over the penis and genitals, and massaged with soot'.[17] In Yorkshire too there used to be all sorts of tricks played on young boys.

> When we went down the pit we all had to go through our initiation ceremony, during which they used to pull our trousers down and examine our little sparrow. The size of that was very important. If we were well endowed, we were looked on with great respect; if we had a poor, weedy little thing, they used to cover it with fat and make fun of us for days, or else they used to paint it and hang a bit of band on it and all sorts of things . . .
>
> Other tricks would be to send a kid to fetch a leather-faced hammer or the sky hook; or else someone would be asked if he had cleaned the windows yet, and if not, would he go round the corner and ask so-and-so for a wash leather for the job.[18]

Mines were always dark, dirty and dusty. A great deal of dust was inevitably created in cutting the coal, getting it down and moving it to the surface. Even in the early nineteenth century the railways used in the deep Northumberland

and Durham pits raised so much dust that the roadways had to be damped down with water to prevent the men from suffocating as they worked. As pits elsewhere went deeper, so more and more dust was created by the spread of mechanical haulage and by what little coal-cutting machinery was introduced. No real success was had in the suppression of this choking, swirling dust until the very end of the century and even then progress was painfully slow. The miner always worked 'enveloped in dust and sweat', and many, like Edmund Stonelake from South Wales, remembered only too vividly 'the foul atmosphere of a nineteenth century coal pit where one's lungs got clogged with dust, and nostrils constantly assailed with foul smells from sweating stinking horses, and perspiring men'.[19]

Little was done either to lighten that 'darkness which might almost be felt' which so frightened all newcomers to the pit.[20] The replacement of naked candles with the less dangerous (but far less powerful) safety lamp may have done something to reduce the risk of explosion. But it did nothing to improve the quality of underground illumination. By the early twentieth century, the electric lighting of the pit bottom and main roadways had become fairly common but elsewhere the use of electric light remained almost unheard of: in 1911 over a million miners shared just 4,298 portable electric lights between them. Away from the main roadways, lighting remained crude. The amount of light falling on the faces of many South Wales steam coal pits at the start of the twentieth century was found to be no more than one seventieth of that produced by a standard candle at a distance of one foot.[21]

Sanitation underground was appalling. Eating, drinking, urinating and defecating all went on side by side. Rats were everywhere 'scurrying all over the place, making sound like a flock of sheep'.[22] The nineteenth-century public may well have been surprised at the number of occasions on which inquests found that an explosion was caused by experienced men 'entering old workings with a lighted candle'. The explanation was simple, if indelicate. Well into the twentieth century 'There was no sanitation whatever, anywhere down below. For a toilet one just went into some old working place which had been abandoned and just made the best of it amid the worst.' Refuge places (in the sides of the roads) 'were the places where indecent miners for convenience deposited their excrement. As refuge places and recesses were used for the storage of timber also it was quite common inadvertently to foul one's hands with dung.' In some districts it was even the custom to drink the stagnant water lying around in the pit. 'Hygiene was unknown and quite foreign, the darkness hiding even the dirty hands with which we handled our food. The proverbial peck of dirt any individual is reputed to consume in the course of a liftetime has no origin nor place in the mine.'[23]

The shallow mines so common in the early nineteenth century were at least cool to work in. But like those near to a coast or a river they were often excessively wet. The collieries sunk along the Tyne Valley at the end of the eighteenth century had to wage an unending campaign against water. By 1850 Friar's Goose colliery, which had been sunk right on the banks of the Tyne, was pumping out a thousand gallons of water every minute of the day and night. In Scotland it was not unusual for miners to 'come up the pit at night drenched to the skin with water, and . . . have to travel a considerable distance home'. Working in wet pits like these is still today an agony of 'flesh driven mad with the chemical-filled water . . . of soaked knee pads rubbing into the bones and wet straps cutting into the skin . . . The slightest cut becomes filled with the salt-like water . . .'[24]

In some respects the working environment deteriorated as the century progressed. All miners knew that the deeper a pit was sunk, the hotter it was to work in. In the early 1880s the heat and stuffiness associated with deep mining was largely confined to the North-East. Thus on a cool day in 1820, the surface temperature at Jarrow colliery was just 46° F; 289 yards down at the pit bottom it was 61°; in the worst ventilated part of the workings it was 75° while in the boiler room it reached a staggering 144° F. With the spread of deep mining to the concealed fields of Lancashire, Yorkshire, the Midlands and South Wales, the heat problem became both more general and more serious. It was estimated that the mean rate of increase in temperature was 1° F for every forty-six feet in depth. So on a warm summer day the temperature in any thousand feet pit was likely to be well up into the nineties.[25]

The severity and discomfort of all work underground was determined a great deal by the thickness of the coal seam. Thin seams were found in every coalfield. In 1841, seams less than eighteen inches high (lower than the seat of a chair) were being worked in Yorkshire, Lancashire, Shropshire, Gloucestershire and Somerset. 'Those who work above ground,' complained one hewer, 'would do well to consider what it would be like to get under the chair, at home, and work with shovel and pick.'[26] Thin seams continued to be worked; indeed one way often used to boost output at the end of the century was to bring into production some of the thinner and less productive seams.[27] The result was that even at the very end of the period almost one third of a sample of forty-three South Wales collieries were found to be working seams that were less than three feet high.[28]

It is true that some of the special sympathy felt for miners in thin seams was misplaced because whatever the thickness of the seam, some work like undercutting (cutting a groove underneath the coal) had always to be done lying, sitting or kneeling down. But almost every other task which the collier and his helpers had to perform was made infinitely more difficult and exhausting for being done in a cramped space. In parts of Northumberland the men in thin seams had to 'go about on their hands and knees', while in Somerset 'some places are scarcely fit for a dog to go in, not being more than from two feet and a-half to a yard in height, and in such places as this many of our boys have to go for twelve hours or more a day ...'.[29] Even in considerably thicker seams, like the valuable Four Feet steam coal of the Rhondda, almost all the hewers' work had to be done on their knees while the putters and drivers, however small, were forced to move round doubled up or stooping. Only in really thick seams like the nine-foot 'Barnsley Bed' in South Yorkshire or the thirty-foot Thick Coal in South Staffordshire were miners ever able to work standing up.

Underground conditions changed then during the course of the nineteenth century, but they varied a great deal from district to district, pit to pit and from seam to seam. It is

difficult therefore to accept wholeheartedly either the pessi-
mistic view that conditions deteriorated as pits went deeper
or the optimistic view which suggests that

> just the contrary was the case. The conditions of the mines
> ... improved with depth from the surface. The seams
> worked in the deeper mines, were usually thicker than in
> the shallow mines. The deep mines were dry and warm;
> and even the presence of the dread fire-damp was not
> without certain compensating advantages, as it necessi-
> tated careful attention being paid to their being thorougly
> ventilated at all times.[30]

The situation is too complex for easy generalisation. Who
can really say if the early nineteenth-century miner working
in a cold, wet, shallow pit infested with choke-damp was any
better off than his grandson choking and gasping in the hot,
dry dusty fire-damp ridden atmosphere of a deep late nineteenth-century colliery?

Not only were conditions in every pit unpleasant, they were
also constantly changing. There was the ever-present possi-
bility of the roof collapsing. The seam might become thinner
or harder; it might begin to contain less and less coal and more
and more stone; or it might disappear altogether. Standardisa-
tion was impossible and in the dark workings close super-
vision almost out of the question. In these circumstances there
was a good deal of quarrelling, bullying and intimidation.
Apparently trivial disputes sometimes had fatal consequences.
A boy of sixteen working in a pit near Bristol in 1825 got
into an argument with another miner over a candle and
stabbed him to death with the candlestick.[31] Pit lads were
notoriously unmanageable and their 'predilection for mis-
chief' well-known.[32] A newcomer to the Hamsteels colliery
near Durham in 1910 'went to the place where putters and
drivers assembled for their baits (lunches) and in so doing
surprised two adolescents, RH and CC, who were on their
backs masturbating in competition to find out which could
produce an orgasm first'.[33] Victimisation was easy. Blacklegs
who carried on working in Northumberland and Durham
after the collapse of the 1844 strike had stones mixed with
their coal, their clothes and tokens stolen and had to face

possible decapitation from ropes tied across the main road-ways at neck height.[34]

Such indiscipline in the face of constantly changing conditions, the indifference of many owners, and the almost criminal neglect of some, combined to make coal-mining not only unpleasant but also exceptionally dangerous, more dangerous than any other occupation with the possible exception of deep-sea fishing.[35] To the natural hazards of underground work had to be added the undoubted fact that only too often familiarity did breed contempt. Time and again miners — particularly the young or the inexperienced — diced with both their own lives and with those of their mates. Like the young Thomas Winfield of Shipley in Derbyshire, they took matches underground. Like Edward Panton of Rotherham, they ignored danger signals. And like James Ball of Pendlebury (who, though working as a hewer, was really a surface tradesman) they refused to set their roof props close enough together.[36] The men firmly believed that 'many accidents were caused by agricultural labourers, brickmakers, and other unqualified men being put into responsible positions in the mines'.[37] It is impossible to compile accurate accident statistics for the first half of the nineteenth century but it does seem that with the expansion of the industry the number of fatalities a year doubled from about 300 to around 600.[38] It is known that from 1850 until World War I an average of more than a thousand miners were killed annually, with 1866 (1,484) and 1913 (1,753) being especially bad years. The vast majority of all accidents — probably more than 95 per cent — took place below ground and a disproportionate number happened to the youngest workers. Thus although boys aged between ten and fifteen constituted only a ninth of the South Wales workforce in the early 1850s, this group suffered more than a fifth of all fatalities.[39]

To most contemporaries, and to most historians since, the problem of industrial safety has been seen almost exlusively in terms of fatal accidents and, in particular, of major disasters. Certainly the names of even a few disasters make sober reading, evoking the memories of untold suffering and despair: New Hartley (204 dead), the Oaks (361 dead), Seaham (164 dead), Clifton Hall (178 dead), West Stanley

(169 dead), Senghenydd (439 dead).[40] The frequency and scale of these catastrophies meant that the dangers of pit work were universally recognised. A spectacular explosion, entombment or inundation, emasculating a complete community, made exciting and popular reading as a glance through the files of any nineteenth-century newspaper will readily confirm. The reports were often harrowing. Sometimes the doomed men were able to leave a last message to their loved ones. Michael Smith, one of the 164 killed in the Seaham explosion of 1880, scratched the following message on his water bottle:

> Dear Margaret there was 40 alltogether at 7 am. Some were singing Hymns, but my thoughts were on my little Michael that him and I would meet in heaven at the same time. Oh Dear Wife, God save you and the children and pray for me Dear Wife Farewell, my last thoughts are about you and the children, be shure and learn the children to pray for me. Oh what an awfull position we are in.[41]

It is almost impossible to overestimate the impact of a large colliery disaster on a small, single-industry mining community like those in parts of Yorkshire, South Wales or Northumberland and Durham. It could destroy virtually a whole generation of men and youths, leaving behind only 'a company of aged men, weak women, and helpless children'.[42] A visitor to Haswell in Co. Durham after the explosion of 1844 had claimed ninety-five lives, found that in the Long Row 'every house save one had its dead. In one house five coffins stood — two on the bed, two on the dresser, and one on the floor.'[43]

Yet horrifying and destructive as these great disasters were, they accounted in fact for but a small proportion of all the lives lost in the coal industry. As the miners and their families realised only too well, the vast majority of deaths were caused not by famous disasters, but by isolated, and therefore almost entirely unpublicised accidents: 'The one widow whose husband as he worked alone in the mine was struck dead by the fall of a stone, the two or three women whose bread-winners met their death through a shaft accident, the aged mother whose lad has been run over in the workings and after a lingering period of pain has left her utterly destitute . . .'[44]

This steady drip-drip of death never rated more than a brief mention in the local paper. Yet behind the sentence or two which did begin to appear as the century wore on, there lay great misery and struggle. These three examples taken at random from the Barnsley press are typical.

> Yesterday afternoon Joseph Roebuck, a miner employed at East Gawber Colliery was killed by a fall of coal. He was so completely covered that it was a considerable time before his body could be got out.

> On Saturday afternoon an inquest was held at the Fox and Hounds, Shafton-two-Gates, near Barnsley, touching the death of Samuel Davis, aged twenty-nine years, a pit sinker, who was killed at the above colliery by a fall of stone on the 18th instant. The jury returned a verdict of "Accidentally killed by a fall of stone."

> On Thursday evening, while two men, named James Priestman and Isaac Jones, were working in number 14 stall, a slip occurred in the roof, and a large quantity of coal and bind falling upon them, they were immediately killed. Priestman was a single man, but Jones has left a wife and several young children.[45]

Such deaths did not make much impact even in mining communities themselves. The Children's Employment Commission was told that around Oldham for instance there were so many people killed in these day-to-day accidents 'that it becomes quite customary to expect such things. The chiefest talk is just at the moment, until the body gets home, and then there is no more talk about it.'[46] Even if the definition of a mining disaster is diluted so as to include any accident claiming more than four lives, disasters never accounted for more than a quarter of total fatalities. So it is important to remember that of the thousand or so miners who died each year at the pit, well over 750 were usually killed in those isolated occurrences which constituted 'Colliery disaster in instalments'.[47]

Even more misleading than the lack of attention paid to these small-scale fatal accidents has been the almost complete neglect of non-fatal injuries.[48] This is hardly surprising for there exist no reliable statistics of either the number of non-fatal accidents occurring in the industry or of the number of working days which were lost as a result. Yet it is possible to assess the incidence of non-fatal accidents with a fair degree of accuracy. It would be misleading to place too much faith in contemporary commentators, many of whom tended to equate the problem of mining accidents exclusively with fatal occurrences. Even the official returns of non-fatal accidents which were made after 1850 were 'of little value in a statistical point of view'.[49] So it is necessary to base estimates of the number of non-fatal accidents on the known number of fatal injuries. A return prepared for the Durham Coal Owners' Association in 1898 reveals that during the previous two years there had been 98.9 non-fatal accidents for every death caused by an accident killing less than five miners. This ratio is confirmed by the extensive experience of the miners' permanent relief funds, friendly societies which, it will be seen later, were designed specifically to insure miners and their families against both fatal and non-fatal accidents. In 1892 the chairman of the finance committee of the Monmouthshire and South Wales fund disclosed that the proportion of non-fatal accidents to fatalities (excluding those caused by accidents claiming five or more lives) occurring to members of the permanent relief societies also remained constant at almost one hundred to one. It is not quite so easy to determine how many working days were lost as a result of these non-fatal accidents. But again the records of the permanent relief fund movement prove a most valuable source. They show that their 167,000 members injured in Lancashire and Yorkshire between 1872 and 1896 were generally away from the pit for about thirty working days. Since there is no reason to suppose that the experience of these two coalfields was in any way exceptional it may be concluded with some confidence that the ratio of non-fatal injuries to deaths arising from accidents claiming less than five lives remained constant at almost exactly one hundred to one and that the average non-fatal accident resulted in the loss of some thirty working days.

Working from the known figures of fatal accidents it is thus possible to compile estimates of the incidence and duration of non-fatal accidents. One important fact emerges immediately:

by the end of the century, if not before, more working days were lost from injuries than as a result of the far better-known strikes and lockouts. Accurate statistics of industrial disputes first became available in the 1890s and in 1896, for example, 1,012,000 days were lost because of disputes in mining and quarrying. Yet in England alone 505 miners died in that year, which means that about 50,500 men received non-fatal injuries entailing the loss of over one and a half million working days. During the following year, when 867,000 working days were lost in English industrial disputes, there were 550 miners killed in small-scale accidents and around 55,000 non-fatal injuries resulting in the loss of about 1,650,000 working days.[50] 'Broken bones in those days,' remembers a South Wales miner, 'were a common occurrence, and were treated with scant respect.'[51]

In considering the impact of industrial accidents on nineteenth and early twentieth-century mining communities the historian must look then beyond the spectacular and well-documented major disaster. He must also consider the single fatality and the great mass of non-fatal accidents, the overwhelming majority of which did not receive so much as a cursory mention in the local paper. The miner was four times as likely to die in an isolated incident — and at least 460 times as likely to be injured — as he ever was to be killed in a great explosion. Or to put it another way, of every thousand miners working in the industry between 1879 and 1890, 0.3 were killed each year in disasters and 1.7 in other fatal accidents, while every year a further 167 had to stay off work with less serious injuries. Between a fifth and a sixth of all underground workers were injured every year.[52]

As well as taking account of these different types of industrial accident, the historian needs to be fully aware of the marked regional variations in the miners' liability to injury. In Yorkshire, the North-East, Lancashire and South Wales a combination of adverse geological conditions and rapid industrial expansion combined at different periods to increase the number of major disasters far above the national average. There were 'gasey' seams in Northumberland, Durham and Lancashire, and seams liable to spontaneous combustion in parts of Staffordshire, in Yorkshire and in the steam coal

districts of South Wales.[53] The natural difficulties of mining
in these coalfields were aggravated as the industry expanded
into deeper and gassier seams. We saw above how the late
eighteenth century witnessed the Northumberland and
Durham coal-owners beginning to move out from the Tyne
Valley. A series of disasters promptly followed: Hepburn,
1805, thirty-five dead; Felling, 1812, ninety-two dead;
Heaton, 1815, seventy-five to ninety dead, and so on . . .[54]
A similar pattern of expansion followed by tragedy recurred
in other coalfields. When the industry in Yorkshire began to
move from the western outcrop to the Barnsley area, the new
area experienced some terrible explosions: Lundhill, 1856,
189 dead; the Oaks, 1866, 361 dead and Swaithe Main, 1875,
143 dead. Lancashire in the 1850s and 1860s saw the deepest
sinkings in the country. Almost inevitably, it seems, such
pioneering technology was followed (between 1869 and 1871)
by a series of ten disasters in which 317 lives were lost.[55]
Also at mid-century the first sinkings were being made to
the fiery steam coals of the Aberdare Valley. There followed
the series of disasters which were to become so horrific a
feature of South Wales colliery life: Cynmer, 1856, 114 dead;
Risca, 1860, 142 dead; Ferndale, 1878, 268 dead . . . right
up to Britain's worst ever mining catastrophe when 439 lives
were lost at Senghenydd in 1913. But while miners working
in South Wales, Lancashire, Northumberland and Durham
were especially at risk from large-scale accidents, those working in the Midlands or the West of Scotland were practically
immune from such disasters.

Equally marked were regional variations in the incidence
of other fatal accidents and of non-fatal injuries. These types
of accidents also resulted to a large extent from the aggrava-
tion of natural geological dangers by working at greater depths
where roof pressures were higher, haulage distances longer
and winding more difficult. So it is not surprising that by
any measure the most dangerous districts in which to work
were West Lancashire, Staffordshire and South Wales. The
safest (and the term is of course strictly relative) were Scotland, South Durham and the Midlands. The contrast between
the high and low risk fields was considerable. The best evidence
suggests that it was at least twice as dangerous to work in

South Wales or Staffordshire as it was in the Midland coalfield of Leicestershire, Warwickshire, Derbyshire and Nottinghamshire.[56]

Mining accidents were always very, very frequent. But this
must not blind the historian to the declining incidence of all
types of accident from the mid-nineteenth century onwards.
Between 1850 and 1914, the likelihood of a miner being
killed in an industrial accident fell almost three and a half
times. It was a substantial improvement and one which contributed largely to the improved conditions of life enjoyed
by the mining community during the second half of the
century. But while the death rate — as opposed to the number
of deaths — did show a marked decline, the growth of the
industry meant that throughout the period large numbers of
men were killed at work. Between 1868 and 1919 a miner
was killed every six hours, seriously injured every two hours,
and injured badly enough to need a week off work every two
or three minutes.[57] The psychological impact of working in
such a dangerous industry was incalculable: 'no man knows
when he leaves his happy fireside in the morning but ere
night he may be carried home a mangled corpse.'[58]

Oh let's not think of tomorrow lest we disappointed be,
Our joys may turn to sorrow as we all may daily see.
Today we may be strong and healthy but soon there comes
    a change,
As we may see from the explosion that has been at Trimdon
    Grange.

Men and boys left home that morning to earn their daily
    bread,
Nor thought before that evening they'd be numbered with
    the dead.
Let's think of Mrs Burnett, once had sons but now had
    none —
By the Trimdon Grange explosion Joseph, George and
    James are gone.

February left behind it what will never be forgot.
Weeping widows, helpless children may be found in many
    a cot.

Now they ask if father's left them, and the mother hangs
her head,
With a weeping widow's feelings tells the child its father's
dead.

God protect the lonely widow and raise each drooping
head,
Be a father to the orphans, never let them cry for bread.
Death will pay us all a visit, they have only gone before,
And we'll meet the Trimdon victims where explosions are
no more.[59]

The discussion of occupational disease in the coal-mining
industry must be far less exact than the foregoing analysis
of industrial accidents. Even the distinction between an accident and a disease and the distinction between a disease caused
by occupation and one merely aggravated by it is by no means
easy to make. But in general 'an accident may be defined as a
particular occurrence causing injury at a particular time
whereas a disease is produced over a long period of time.'[60]
Minor ailments were accepted as part and parcel of pit life
and went almost completely unrecorded. It is certain though
that the punishing nature of mining work, like any other hard
physical labour, caused many strains and ruptures, that working in the damp caused boils and rheumatism and that working in gassy seams often resulted in headaches and loss of
appetite. If few miners died from rheumatism or rupture,
many died with them.[61] Also common was the group of dis-
abilities known as beat hand, beat knee and beat elbow,
'the plague of the collier working in the cramped quarters
of the thin seam. It comes from the constant chaffing of
the stone floor against the knees . . . and elbows, or the wear
of the shovel against the hands. This results in the festering
of the joint or limb and a huge inflammation area develops.'[62]
Again there are no reliable statistics of these crippling afflic-
tions. But pit work changed very little during this period and
there is certainly no reason to believe that they became less
common as the century progressed.[63]

Prevalent too were lung diseases such as 'miner's asthma',
the crippling breathlessness which we now call pneumoconiosis. Every underground workman inhaled dust and im-

pure air so that 'Ultimately his lungs are loaded with black matter, solid or fluid, like printers' ink, or common ink, or lamp black, or charcoal powder, all insoluble and tasteless.'[64] Miner's asthma seldom affected young men but by early middle-age almost all underground workers were showing symptoms: 'there are few young men above the age of twenty-five who are quite free from pectoral disease in some shape or other; and above the age of thirty-five there are not ten per cent who do not suffer more or less from asthmatic disease. Above the age of forty almost ALL miners are the subjects of chronic bronchitis and asthma.'[65] So although itself rarely a killer, miner's asthma made the lungs unable to cope with even mild attacks of bronchitis or pneumonia which accordingly went on to cripple and kill many wor-kers.[66] One of the two chief causes of miner's asthma, the presence of coal dust underground, did not get any better as the period progressed. But the other major cause, poor ven-tilation, did improve so that the disease did become markedly less widespread during the second half of the century.[67] But still on the eve of World War I every mining community had its 'men with the deadly dust in their lungs, waiting only for death'.[68]

An even more serious disease of the lungs was 'miner's phthisis' or 'black spit', the present day silicosis. It was the cause, remarked one doctor with collier patients, of 'a terrible death, and one never to be forgotten'.[69] Its victims were those who inhaled a stone dust found in parts of Somerset, South Wales and particularly the East of Scotland. So mercifully it was highly localised and it too was to become less common as improvements were made to underground systems of ventila-tion.[70]

Probably the best known of all the occupational diseases of the miner was miner's nystagmus whose 'chief symptom and physical sign is a rotary oscillation of the eyeballs, which prevents the miner from accurately fixing anything towards which his vision is directed'. Other tell-tale signs included headaches, giddiness, 'night blindness' and photophobia (dread of the light). Except in the most severe cases, however, a complete recovery could be had within a year or two of leaving the pit.[71] Nystagmus was first indentified in Belgium in the early 1860s and became generally recognised in this country from about 1875. There was a protracted debate between Josiah Court and Simeon Snell as to whether the disease was caused by poor illumination or by the upward direction of vision arising from the miners' working position. Eventually Snell's deficient light hypothesis prevailed.[72] Court and Snell clashed again though over the incidence of the disease. But in a careful study published in 1912, Dr T. L. Llewellyn found that the complaint was six times as common in pits worked with safety lamps as it was in pits worked with naked lights. By this date between 0.5 and 2 per cent of all underground miners, that is nearly 10,000 men and boys, were unable to work on account of nystagmus.[73] Here then was a disease which, though unknown at the beginning of the nineteenth century, increased in virulence with the spread of the safety lamp until it was well established at its end. In the early years of this century there were many men like this fifty eight year old collier who recalled how

> Up to the last two years before I failed I had no trouble with my eyes and always earned good money. During the last two years my eyes got weak, but I struggled on, hoping things would mend. I lost days and days, and on times a week. At the time it was not safe for me to go to the face without the help of another man. I could not recognise anybody, and had to walk in with my lamp held behind my back. I could always tell by the sound if it was safe. My wages fell to a pound a week, and the manager stopped me at last and told me that it was not safe to allow me to work any longer. If I could only have known before, I might have saved my eyes.[74]

Among the less common occupational diseases of the miners may be counted blood-poisoning, glanders and ankylosto-miosis. Pony drivers were liable to catch glanders from their charges: in 1862 for example a boy working underground at the Camerton coal works in Cumberland died after catching the disease from the horses.[75] Blood-poisoning killed a hand-ful of miners nearly every year, often as the result of wearing really filthy clothes.[76] Ankylostomiosis, also known as worm disease or miner's anaemia, was first identified in 1902 and thereafter efforts were made to improve sanitation facilities and to acquaint miners with the dangers of the disease.[77]

The history of miners' diseases is still to be written. It is known, however, that while the danger from disease varied from coalfield to coalfield and from pit to pit, there was some general improvement as the century wore on. Yet the threat of death and disablement by accident or disease remained more common in mining communities than elsewhere.[78] A South Staffordshire man remembers that his brother was killed underground and that his father broke both his thighs and later developed nystagmus.[79] So Arthur Horner's experience, tragic and horrifying though it was, was not that unusual.

> One Sunday, somebody left a ventilator door open in the Glynmeal Level, and my grandfather, going in at evening to examine the pit, was blown to pieces by an explosion. They collected his remains with a rake, and brought him home in a sack ... My uncle worked in the mines until his back was broken by a fall of roof. Another uncle went nearly blind with nystagmus. I learned very early that there was blood on the coal.[80]

What then of the jobs which the miners had to perform in this constantly changing and threatening environment? The term underground worker is sometimes used as if it is synonymous with face-worker. It is not. A good half of the underground labour force was employed not to cut the coal, but to service those who did.[81] It was their job to move the coal from the face to the shaft bottom and to maintain the underground workings in safe and workable condition. They were known as oncost workers or as day workers because of the way in which they were paid. Like surfacemen, oncost workers were not regarded by the hewers as real miners. They were children and youths, sometimes even women and girls. They were serving their apprenticeship in underground work. They were not like the hewers who had to pit their strength and skill against the coal itself. Face-workers always had a remarkable blind spot where oncost and surface men were concerned. In 1865 Robert Woodward, a hewer from the Haydock colliery in Lancashire, complained to a parliamentary Select Committee that one of the officials where he worked 'had never done a day's work at all in the pit'. What he actually meant to say, he explained later, what that the official in question had only ever worked as a daywageman.[82]

There were always a number of odd jobs that any child could do when he first went below. In Durham in the 1840s young Neddy Rymer was put to work 'helping-up, switch keeping, lamp-oiling, way cleaning'.[83] Later in the century as haulage methods improved the new recruit might also be asked to help operate a steam engine or a self-acting plane (a haulage system worked by gravity).[84] Very young children might be eased in more gently. William Baxendale was taken down the Black Rod colliery near Wigan before he was five years old but he was allowed to knock 'about as an errand boy . . . for a time' to get him used to the pit.[85]

Still for many the first taste of pit life was trapping, the opening and closing of ventilation doors to allow men and coal to pass through without disturbing the current of air through the workings.[86] Some children like William Baxendale did go underground at four years old but even in the early nineteenth century the majority of trappers did not start until they were about eight years old. Even so, in 1841 there were some 5,000 children between five and ten years old working underground, and most of them were probably trappers. Slowly the starting age was forced up until by 1914 no boy under fourteen was allowed to go below in any capacity whatsoever.[87] George Parkinson went down when he was nine.

> A few hundred yards from the shaft bottom brought us to a large trap-door, about six feet square, closing the whole avenue . . . my father set to work to make a trapper's hole behind the props, in which I might sit safely and comfortably. After hewing out a good shelter for me he put a nail in the door, to which he fastened my door-string; attaching the other end of it to a nail in a prop where I sat, so that I could pull the door open when the horse and wagons were coming through without exposing myself to danger.[88]

To 'sit crouching in a dark, damp hole behind a door' was not hard work. But it was boring. Anybody with young children will readily imagine how long a ten or twelve hour shift must have seemed to a small boy or girl. The child was 'imprisoned there, just the same as if he was in a cell in a gaol'.[89] In slack periods he might be left alone for a long time. George Parkinson's 'candle went out, and, all alone in the darkness . . ., I sat in my hole afraid to breathe'.[90] In one of these quiet moments at the Plymouth collieries near Merthyr Tydfil, six year old Mary Davis fell asleep beside her door: 'She said the rats, or some one, had run away with her bread and cheese, so she went to sleep.'[91] The only break in the monotony came when somebody came through the door. But even this was a mixed blessing for the older boys liked nothing better than to bait the trappers. They bullied them, stole their food and put out their candles. The young Rymer had to mind two doors on an incline and whenever they could 'the drivers flung coal and shouted to frighten me as they went to and fro with the horses and tubs.'[92] The trapper's job may not have been hard, but it was totally unsuitable for young children. According to one retired South Wales collier trapping was "The worst job going'.[93]

> . . . . . . the trapper,
> That's the name they give the door boy,
> And it's a queer job.
> A'l on me tod. . . .[94]

After a couple of years or so trapping the young miner generally moved on to some type of haulage work. As a hand putter, he (and sometimes she) had to take empty tubs to four or five hewers and then bring the full ones back to the nearest main roadway or, as in the early part of the century, right to the pit bottom. The putter could move round the workings; he could bully the trappers and check the hewers. No longer was he the baby of the pit. The putter's work was hard, probably the hardest done by any miner except the hewer himself. The drawings commissioned by the Children's Employment Commission did not exaggerate. All over the country in the 1840s (and beyond) children and youths were struggling to move huge containers of coal, sometimes weighing as much as a quarter of a ton. As one fourteen year old

Welsh putter commented, 'it is very hard work indeed, it is too hard for such lads as we, for we work like little horses.'[95]

> A'm on me way inbye, and working as a putter,
> Shifting tubs from flats and partings to the hewers.
> They're light on the flat, but coming back.
> Yi'd swear each tub weighed a thousand tons.[96]

In the first half of the century a good number of women and girls worked as hand putters in the small, technically backward parts of Lancashire, Gloucestershire, the West Midlands, Pembrokeshire and the East of Scotland.[97] In the steep seams of Pembrokeshire they used a windlass and chain to raise coal from the workings to the level and in this way a pair of women might raise as much as eighty tons in one eight to ten hour shift.[98] In the Forest of Dean and throughout the East of Scotland women worked as 'bearers', carrying well over a hundredweight of coal through the workings and then 'with weary steps and slow . . . halting occasionally to draw breath' up narrow stairs or ladders 'till they arrive at the hill or pit top . . .'[99] But by mid-century it was exceptionally rare to find female putters, or indeed any women and girls, working underground. And by mid-century ponies — and even some engines — were being used to move coal along the main roadways. These developments might lead one to suppose that by this time putting was no longer the exhausting job which it had been before. Certainly these haulage improvements meant that the tubs had only to be moved to the nearest main roadway rather than all the way to the pit bottom and this meant in turn that fewer putters were required. But one would be hard pressed to argue that technological change really did much to improve the lot of those putters who continued at work. Putting distances were reduced, it is true. But to make up for this each putter was given more places to service. And distances were still long enough. As pits got older and faces retreated from the main roads, the distance which the coal had to be hand-putted could easily stretch up to a quarter of a mile.[100] Nor did the introduction of ponies and engines underground do anything to lessen the constant pressure on the putters because 'Every other man's living depends on those employed

to push and drag the tubs from the coal face to the "flats".'
So the hewers' object was always the same, by bribery or by
intimidation to get as much work as possible out of their
putters. 'Of all the kinds of work in a mine putting is the
hardest. It is surrounded by more dangers and difficulties
than can be conceived by those unacquainted with the labour
of the miner.'[101] Here a newcomer to the industry describes
his first putting shift in the Fife coalfield at the beginning of
this century.

An empty hutch weighs nearly five hundred pounds.
In appearance it is like a small railway coal wagon. An
average load is from half a ton to twelve hundred pounds
of coal. Fourteen or fifteen hundred pounds is a fairish
load for a muscular man.

I started on my first trip. First a dead level, followed by
a slight rise, another short level then an abrupt fall, not
sufficiently abrupt to be characterised as steep but so
inclined that it would have sent an unrestrained or un-
balanced car forward at so bounding a rate that it would
have left the rails at the first bend, of which there were
several. It took every particle of my strength to mount
the first incline and with a sense of relief I felt the forward
end drop as I gripped the other to hold it back. An uneven
bit of rock caused my foot to trip over a sleeper, the hutch
gained in speed till I was jerked off my feet. The hot air
cooled as I was dragged on with quickly increasing speed,
faster and faster. I struggled with might and main to hold
back, but it was useless. The thing had gained a terrible
headway, by great leaps and bounds I went stumbling into
the nothingness ahead at a mad pace; my lamp was blown
out before twenty yards had been covered and there flashed
a picture of the one hundred and sixty or more yards to
go; clinging desperately as if for my life, my weight hanging
all too loose on the end of the runaway hutch barely
balanced it to the rails. If I rose to three quarters my
height I knew I would crash against the stone roof with
terrific force, if I let go, a hard tumble would be inevitable.
Not knowing what was in front was terrible, and the
thought of reaching the end of the level where men, ponies

and long races were passing with every few seconds, was sickening, as with crouching leaps we — the hutch and I — went careering on, till with a joyous thrill I found it coming more and more under my control and at last it rolled gently on to the switch as if the whole run had been just as usual. Every muscle in my body felt pulled out and my tongue was cleaving to the roof of my mouth like dry leather. There was naught to do but relight my lamp, get behind an empty hutch, and laboriously push it back to the face. How my legs stiffened and ached under the strain! My breath came in wheezes and every pore seemed a tiny spring. With greater determination I started upon the second trip, when to my unaffected horror it was the same madcap rush over again, only worse. My fingers would not act, my strength seemed to be running like the sweat from every limb. How the hutch kept the rails throughout that breathless, perilous run I shall never know. The heat was cruel. With violently trembling hands I grasped my flask and swallowed a mouthful of tea, lukewarm but refreshing. My lips were like blotting paper.

Until now my mate, a broad shouldered fellow with herculean biceps and chest had not spoken a word, but as he passed he said lightly:

"After my first shift on this job I thought I was dead."[102]

After two or three years the putter was generally moved on to pony driving. This was progress, 'a proud elevation', the next best thing to being a hewer.[103] Driving was far easier than putting and much more exciting than trapping. Drivers, or hauliers as they were called in some areas, were in their mid-teens with four or five years underground work behind them. They had the freedom of the pit and could not wait to start hewing. They caused no end of trouble.[104]

Though the boys would never appreciate the fact, driving was very dangerous. They were at risk from the tubs, from the ponies, from low rooves and from uneven walls. The Durham miners' leader John Wilson recalls a narrow escape he had at Ludworth colliery.

There had been an incline shaft driven, and for some

purpose the drift was laid with bricks in the bottom of the ways. I was attempting to take a tub down the drift, and thought I could manage by going down before it, and pushing against it. I had barely got started when my feet slipped and got under the tub. I was thus pushed down, but my legs acted as a drag. I was being taken down at an unpleasantly quick rate, and the great danger lay in my being dashed against some tubs which were standing at the bottom. My shouts brought to my aid a waggonwayman, and he was able to stop me just in time. If he had been a moment or two longer I should certainly have been crushed to death. By his timely intervention the only damage done was a few days' soreness for me, and the outlay by my kind friends for a new pair of trousers, for the ones I started with from the top were partly left in the drift, and the remnant in rags hanging to me.[105]

The drivers were also at risk from the pit officials. Their general misbehaviour meant that they invariably took the blame when anything went wrong. If a tub ran off the rails, the driver was blamed. If a driver was caught taking a rest, he was blamed for that too.[106] In every pit the drivers, and to a lesser extent the putters, seemed to bear the brunt of whatever violence was going.[107]

Most miners started underground young, passing 'from childhood to manhood through the ordinary curriculum of the . . . pitboy's lot'.[108] A child going down the pits at eight, graduating to hewing at eighteen or nineteen, and stopping work at fifty or so would spend a quarter of his working life as an oncost worker. This experience, at a most formative stage of his life, shaped the miner's entire physical and emotional development. By the time he began hewing the miner was already likely to be strong, aggressive, suspicious, obdurate and obsessed with his job.[109]

Eventually, at the age of eighteen or twenty, the great day came. At first he would probably work as a 'spare' man filling in whenever a regular hewer was away. Finally he would be given his own place to share with one or two mates, where he could hew a regular shift.[110] 'Like an apprentice completing his "time", so the "putter", . . . becoming a hewer, has reached his highest level, and in the old pit phrase, "He's now a man for hissel".[111] He was one of the élite group of face-workers — 'the centre of the mining system' — which made up about half the underground labour force.[112] So out of every ten miners in the industry, from three to four (by far the largest single group) were actually employed to get the coal from the seam.

There were considerable variations in the ways that hewing was organised. At the beginning of the century the usual system was the bord and pillar whereby the hewer worked on his own: 'Each collier had his own working place, in much the same way as an allotment holder has his own plot. He worked his plot in his own way without interference from anyone . . .'[113] Later it became more common in the Midlands, Lancashire and South Wales to adopt the longwall system (or its variants) under which a number of hewers worked together along the same coal face.[114] Whether or not the introduction of longwall was 'a step towards the factory organisation of labour', the actual job of hewing changed very little.[115]

> The coal is first holed or undercut with a special tool known as a pick or mandril. That is, a groove one or two feet in depth is cut either in the lowest part of the coal or in the underlying fire-clay. The mass of coal is supported during the operation by sprags or props. When the holing process is completed several sprags are withdrawn and if the immense downward pressure of the overlying strata proves insufficient to break down this coal, wedges are driven in at the top of the seam or [after about 1830] explosives are used in non-gassy mines.[116]

Coal-mining was always 'a pick and shovel industry'.[117] Even in 1913 less than eight per cent of British coal was machine cut. Mechanisation was most popular in old established districts like West Yorkshire, Durham and Scotland where the most easily worked coal was nearing exhaustion. In 1913 twice as much coal was mechanically mined in West Yorkshire as in the more recently developed South Yorkshire field. Indeed in Scotland as much as a fifth of all the coal brought to bank was being cut by machine.[118] Very slowly then the hewer's muscle power was starting to give way to the electricity or the compressed air of the coal-cutting machine. But before World War I nearly all British coal was got by the primitive, back-breaking work of the hewer. It was this capacity for prodigious physical exertion that made the coal hewer, in the words of one ex-Durham miner, 'the highest unofficial position attainable at the cost of the hardest form of mining-labour known'.[119] The closer the miner worked to the coal face, the higher the physical demands, earnings – and hence status. But mere physical strength was not enough. If the face-worker and his mates were to make money and stay alive, he needed, if not a high degree of manipulative skill, then a sure understanding of the dangers of underground work. This sixth, or pit, sense could be obtained only by long experience. Here is one reason why the miner's apprenticeship was so long for as Thomas Burt explained, though 'It does not at all make him more skilled at man's work; it makes him acquainted with the mine.'

> My father always used to say
> pit work's more than hewing
> You've got to coax the coal along
> And not be riving and tewing. [pushing and pulling]

> Noo the deputy crawls frae flat to flat
> While the putter rams the chummins. [empty tubs]
> But the man at the face has to know his place
> Like a mother knows her young 'uns.[120]

Everybody is agreed on the enormous physical effort which hewing demanded. The question inevitably arises: how could the hewer keep up this effort, hour after hour, day after day, year in and year out. The answer is simple. He did not. In the first place, hewers rarely worked the excessively long hours which we tend to associate with nineteenth-century pit work. Stories of miners never seeing the light of day were not told about face-workers. Unlike the surface and oncost workers, hewers were paid by the piece and rarely had to work fixed hours. But they did tend to work regular hours. They were tied to set periods for ascending and descending the shaft, they had to cut a certain minimum each shift and as will be seen later, they liked to eat with their families – or drink with their mates – at regular times. Yet it is still not easy to be very specific about the length of the hewer's shift. The evidence is fragmentary, unreliable and difficult to interpret.[121] It is clear however that face-workers' hours were relatively short. The working-day remained fairly constant throughout the first fifty years of the period: seven to eight hours in the North-East; eight to ten in the South-West; ten or eleven in Yorkshire; about twelve, the average, in Lancashire, the Midlands, the West Midlands and South Wales; and up to as high as fourteen or even more in parts of Scotland. After 1850 the length of the shift decreased fairly steadily. The Fife and Clackmannanshire men won their eight-hour day in 1870, and never lost it, while the Black Country miners got theirs two years later, although here it proved far more difficult to keep. By the start of this century no hewer was underground for as long as ten hours a day. Hours were longest in Lancashire and North Wales. Elsewhere, an eight and a half hour shift was usual in Yorkshire, a seven and half hour day in Northumberland and Cumberland, and a working day of under seven hours in County Durham. Finally in 1908, after forty years' agitation, underground workers won a nominal eight-hour day, though in practice it is true it remained nearer to nine.[122] The twelve hour shift, typical in 1850, was unknown fifty years later when most hewers were underground for under nine hours a day. These were not long hours by contemporary standards. At the end of the century many farm labourers, for example, were still working a twelve hour day and longer.[123] Hewers' hours were also shorter, and fell faster, than those of the coal industry's surface and oncost workers. Hewers' hours were long, but not so long as is generally supposed.

Here then was one reason that face-workers were able to continue at their apparently impossible job. But it would be naïve to suppose that they always worked their hardest during the hours that they were at the face. If shortage of tubs, mechanical failure or the inefficiency of the putters disrupted the supply of containers, then the hewer had no choice but to stop work. In South Yorkshire, for instance, engines and self-acting planes frequently broke down with the result that hewing had to stop for an hour or two.[123] Far more important was the fact that hewers were paid according to how much coal they produced and so were to some extent able to decide how hard they wanted to work. In Scotland the 'Darg' was introduced time and again in an effort to restrict output and so, it was hoped, to increase the level of wages. When the 'Darg' was adopted in the West of Scotland in 1879, the length of the working day fell and production dropped by an average of 20 per cent.[124] There is a great deal of other evidence to confirm that very often the hewers did not work to their full capacity. They were nearly always able to increase output when they had a mind to, after a strike or before a holiday. 'The week before Christmas or any other holiday was known as Bull week. The miners tried to fill more coal that week to earn extra money for the holidays.'[125] And output per hewer was always highest at the beginning and end of the pay week or fortnight. The middle of the pay cycle was a time for taking it (relatively) easy. At one colliery in the East of Scotland in 1820, for example, output per hewer on any Wednesday was found to be over a third down on the figure for Saturdays.[126] It is well known that absenteeism invariably increased when piece rates were high during the upswing of a trade cycle. What is less widely understood is that these same high piece rates also encouraged hewers to work less hard when they were at the face. Thus in the boom of 1900 there were constant — and convincing — complaints from owners in many parts of the country that it was taking the men five days to do the work which they had previously done in three.[127]

That the hewer was able to continue at his exhausting, debilitating job year after year was due in part then to the comparatively short hours which he had to put in and to his ability to determine the pace and intensity at which he worked. But there is no doubt that the hewer's industrial longevity was also due in large measure to the number of shifts which he either chose or was forced to miss. Voluntary absenteeism was traditional in coal-mining. It was specially noticeable among miners with working wives and in periods of prosperity when most hewers could earn their money comparatively easily. In the Forest of Dean in 1872

The mouth of every manager is full of complaints respecting the irregularity of the men, and the repeated advances of 10 per cent. that have been conceded during the last few months, so far from promoting the industry and comfort of the miners, appears to have had an opposite effect, for whereas they used to be satisfied with "Saint Mondays", Tuesday and Wednesday are now habitually canonised.[128]

Special occasions always merited a day off. In the early nineteenth century, Northumberland and Durham pitmen stayed at home when the first cuckoo was heard and called a 'gaudy day' to celebrate an engagement.[129] Trade depression was no bar. Even in the summer of 1878 nearly all the pits around Nottingham came to a standstill when the Prince and Princess of Wales visited the city to open the castle museum.[130] Race meetings, elections, displays, galas, fairs and feasts always interfered with production – 'Colliers,' the *Colliery Guardian* complained, 'with the singular improvidence that characterises them, will have their holiday.'[131] So bad was the situation around Leeds that in 1879 a conference was called to recommend the replacement of days off for local feasts by a fixed fifteen days holiday a year (three at Christmas, four at Easter, two at Whitsun and six during Leeds Summer Fair).[132] But little came of the plan. A generation later the *Colliery Guardian* still found that attendance in Yorkshire pits was 'reduced almost weekly by the local feasts, which are held from June to October'.[133]

Miners were among those relatively well-paid workmen who were in the habit of taking off 'St Monday' after their weekly or fortnightly pay. In parts of Lancashire in 1853, for example, 'very many' men were idle every Monday and 'nearly all the rest' every other Monday.[134] But what is not so widely appreciated is the extent to which miners – and especially the hewers – continued to keep St Monday well into the present century. In 1900 there was still a good deal of absenteeism on Mondays and Tuesdays at more than one Staffordshire colliery. But what is most significant is that almost all those who stayed away were the best paid miners, the hewers.[135] It has been seen above that efforts to replace local feasts with fixed holidays can hardly be counted a success. Yet it is important to understand that Christmas, New Year, Easter and Whitsun *were* celebrated, not instead of other local holidays but in addition to them. Nothing could get the miner, especially in Scotland, to work at New Year. Right in the middle of one of the worst depressions of the century, the Fife and Clackmannanshire miners chose to usher in 1879 with an unusually long nine-day break.[136] Easter and Whitsun were also celebrated, particularly in technologically backward districts such as the South-West and South Staffordshire. 'Probably in none of the coal and iron centres, the kingdom through, do the workmen more religiously keep Whitsuntide than in . . . South Staffordshire.'[137]

Voluntary absenteeism was one important factor in allowing the hewers — and to a lesser extent the surface and oncost workers — to continue working at the pit. But it is not easy to assess precisely how important it was. Fortunately, however, some owners in Durham and South Wales did keep attendance records and these show that the hewers' voluntary absenteeism fluctuated between about three and 12 per cent, according to prevailing wage rates and the state of the trade.[138] So even when work was short and wages low, hewers chose to take off about one shift in thirty. When work was plentiful and wages high they would probably decide to stay at home one shift a fortnight.[139] It is usual to see this large-scale voluntary absenteeism in terms of the miner's overall lifestyle, of the low level of his basic needs. As soon as the miner has attained a subsistence level it is argued, 'the first step towards improvement of welfare is not in the direction of goods, food or clothing, but towards leisure . . .'[140] But surely it is not too fanciful to see this large-scale voluntary absenteeism in another light, to detect in it the miner's — and particularly the hewer's — unconscious desire to prolong his underground life. In this sense voluntary absenteeism was not irresponsible. Rather it was a precaution, a primitive form of self-defence.[141]

Another, and much publicised, explanation of the hewer's ability to continue at his murderously demanding job lay in the amount of time lost on account of industrial disputes. Indeed some historians of the nineteenth-century coal industry and miners' unions seem to see developments largely in terms of strikes and lock-outs and play-days.[142] Certainly there can have been few miners who had never been on strike and in some districts at certain times a great deal of time was lost. But the amount of time lost as a result of strikes can easily be exaggerated. It has already been shown in this chapter that in a normal year like 1897 some 867,000 working days were lost by industrial disputes. It sounds a huge number. But in fact it works out to well under one and a half days for every man, woman and child employed in the industry. So even if hewers were twice as likely to become involved in disputes as the rest of the work force, they would still have lost no more than three turns during the course of the entire year. Far more important was the miner's liability to injury and disease. In an average year towards the end of the century, during which a hewer could expect to be in dispute for perhaps three shifts, he would probably lose about five days from pit accidents, probably as many again from industrial diseases, an unknown number from everyday illnesses and at least one day when the pit was closed as a mark of respect for a dead colleague.[143] By his calculation accident and illness together forced the hewer to miss upwards of fifteen shifts a year, a figure which accords well with contemporary estimates of involuntary absenteeism.[144] The conslusion is clear. While strikes and lock-outs did of course play some part in reducing the amount of time that the hewer spent at the coal face, they caused far less absenteeism than sickness and injury, and far less than a reading of the standard miners' histories might lead one to believe.

Probably most crucial of all in reducing the physical demands on the hewer were recurrent periods of enforced unemployment. Pits closed down more often than most people imagine. Between 1854 and 1870 over two-thirds of the colliery concerns operating in the West of Scotland went out of business.[145] Some never really got going. John Hippisley's pit at Clutton in Somerset was forced to close in 1858 because his attempt to find coal had been only partly successful.[146] Others were shut by serious accidents. When the Irish Sea poured into the Union and Lady Isabella pits at Workington in 1837, twenty-five men and two boys were killed – and 300 more laid off.[147] The flooding of Wrexham colliery in 1872 put several hundred men out of work while in the following year 400 men and boys in Derbyshire were laid off for a fortnight following a fire at the Locksford colliery outside Chesterfield.[148] It was during the downswing of the trade cycle, especially in the late 1870s, that most collieries were forced to close. A couple of examples must suffice. Outside Newcastle the Gosforth colliery closed following repeated hints from the undermanager and his officials while in Lanarkshire the Clyde Coal Company closed a one year old pit at Hamilton, a decision which immediately threw 150 colliers and oncost men out of work.[149]

But far more common than full-scale unemployment in the modern sense of the word, was underemployment. Indeed short-time working was one of the outstanding characteristics of the nineteenth-century coal industry. House coal pits were often able to work only a couple of days a week in the summer.[150] And any pit could be stopped by the vagaries of the weather. South Wales pits moving their coal out of Newport in the days of sail regularly lost ten days' work a month from adverse winds and were often unable to work at all during March and April.[151] As might be expected, the most widespread and serious short-time working was brought about by the vicissitudes of the trade cycle. Again there is space for only a few examples. During the 1863 depression many Durham miners were working only three and a half days a week.[152] The effects of the slump of the late 1870s and early 1880s are rather better documented. Nottinghamshire miners were 'only working about three days per week on an average . . .'[153] and at the Shire Moor colliery in Northumberland the men were paid for only seventy-five days between July 1880 and January 1881, an average of no more than two and a half shifts a week.[154] Every downswing in the trade cycle brought the same difficulty. The North Wales Miners' Association was still complaining in 1904 that thousands of its members were working only three days a week, while hundreds more had no work at all.[155]

Face-workers (and other miners to a lesser degree) therefore missed a good number of shifts either from choice, from accident and illness or from industrial disputes. For most of the nineteenth century it is almost impossible to gauge accurately the extent of this absenteeism. But there do exist reliable, if scattered, statistics from the last third of the period. In the early years of this century both the Ebbw Vale and the Powell Duffryn collieries suffered 11 per cent absenteeism among their hewers while the Durham Coalowners' Association produced figures showing that between 1879 and 1907 hewer absenteeism in that coalfield averaged 8.6 per cent.[156] It is not unreasonable to believe therefore that towards the end of this period absenteeism among hewers hovered around the nine to ten per cent mark. Since there is no strong reason to suppose that it was either higher or lower in the first three quarters of the nineteenth century, it may be concluded that for one reason or another most face-workers were in the habit of missing one shift out of every ten that they might have worked.

When there is added to this the hewer's relatively short hours, his ability to decide how hard to work and the large number of days that the pits were stopped, it becomes easier to understand how he was able to continue with his job. This is not of course to deny that hewing was exhausting, back-breaking work. Rather it is an attempt to explain how this awful job could be done by anybody, even by men as tough and resilient as the hewers.

When the hewer grew too old, too weak, too ill or just too tired for face-work, he had few options open to him. Some hewers went back to oncost work, perhaps to one of the more skilled jobs like keeping the roadways dry and in good repair or supervising work at the pit bottom.[157] A few were promoted to firemen, becoming responsible for checking the safety of the workings before each shift began. It was less well paid then hewing but it was easier and the fireman always 'reckons that when he becomes a fireman, he is more advanced towards a managership'.[158] But the growing sophistication of mining meant that there was really very little opportunity for working miners to move into even the most junior management ranks. Most ex-hewers were forced to spend their declining years on the surface, back with the women and children among whom they had started their mining careers. It was a cheerless prospect.

Oh, my name is Geordie Black, I'm getting very old,
I've hewed tons of coal in my time,
And when I was young I could either put or hew,
Out of other lads I always took the shine.
I'm gannin' down the hill, I cannot use the pick
The master has no pity on old bones.
I'm noo on the bank, I pass my time away,
Among the bits of lads with picking out the stones.[159]

3

The Miner's Earnings

Many economic and labour historians, their eyes firmly fixed on price lists, wage rates, trends and averages, have forgotten to go with the miner on pay-day, look at his pay-note and seen how much he actually brought home. They have confused wage rates with earnings, assuming that if they could only rediscover the multiplicity of local — and often verbal — piece-rate and day-wage agreements, they would be able to find out how much miners in the last century were earning. But of course earnings never depended solely on the wage rate. Every miner's earnings depended on the number of shifts which he was able and willing to put in while hewers' and putters' earnings also varied according to geological conditions, management efficiency and the intensity with which they worked.[1] Not only have historians tended to confuse wage rates with earnings; some have also lumped together all miners as if they were earning a broadly similar wage.[2] Nothing, it will be seen, could be further from the truth. Indeed few generalisations about wages, however restricted and hedged about with qualifications, are likely to be of much value in an industry with such large differentials between its best and its worst paid workers.

Even if the historian manages to avoid the twin pitfalls of confusing wage rates with earnings and of lumping together all miners as if they were one, the discussion of miners' wages remains beset with difficulties. No miner ever knew what his next pay-day would bring. Wages varied from week to week, month to month and from year to year. They varied, too, from pit to pit and from district to district. Some miners received binding money, free housing, rent allowances, concessionary coal, cheap firewood, free beer and gifts on special

64

occasions. Many, on the other hand, were made to shop in a company store, subscribe to a pit insurance fund or pay fines for 'unsatisfactory' work.

Any historian who takes the trouble to look at the nineteenth-century miner's take-home pay will see that the outstanding characteristic of the wages system was its uncertainty. The miner lived on a roller-coaster. Even if he stayed in the same job at the same pit and there were no changes in wage rates, the miner could never know what he would be earning the following week, let alone the following year. He could be called out on strike. He could get injured. His rheumatism might grow worse or his nystagmus become unbearable. The pit might close, lay him off or go on short time. When colliery owners went bankrupt miners were often never paid.[3] And so common was short time working that in many villages it was the custom in the evening to run up a flag or send round a caller to announce whether or not the pit would be working the following day.[4]

The hewer could never be sure what stoppages would be kept from his earnings. He knew of course whether his employer deducted money for the pit club or charged (often extortionately) for the tools, candles and powder which he used underground. But he did not know when fines would be imposed on him for 'unsatisfactory' work or for 'offences' committed on the surface. Throughout the first half of the century tubs were confiscated if they were found to contain either too little coal or too large a proportion of dirt. Where miners lived in colliery houses they could also be fined for offences as trivial as keeping dogs, cows, pigs, donkeys or even pigeons.[5] The effect of these large, and often unexpected, deductions on the family economy could be very serious. In South Wales Thursday night was '"jib night" at the collieries. This term is supposed to refer to the men's faces when they get the docket which shows how much pay they should draw on the morrow.'[6] In South Yorkshire a miners' leader claimed to 'have seen colliers on a pay night gaze with astonishment, as though they did not know how it had happened, at the trifle left out of what was apparently a large sum, drawn at the office'.[7]

Seasonal fluctuations were at least predictable. Any miner

working in a house coal-pit could expect to be sacked or, more likely, be put on short time with the approach of summer. Year after year at one Lanarkshire colliery, for instance, a fifth of the workforce was laid off every spring.[8] In Cumberland, Somerset and the Forest of Dean house coalminers also used to have their wage rates reduced by between two and a half and five per cent during the summer months.[9]

Less predictable and more serious were the difficulties brought about by the fluctuations of the trade cycle. In good years both piece-rates and day-wages were relatively high and there was an abundance of work. In bad years, on the other hand, wage rates were low and there was little work to be had. Demand and prices — and hence wages — were especially volatile in the exporting coalfields of South Wales, Scotland and the North East of England. In 1822 South Wales hewers were earning less than 2s. 6d. (13 pence) a shift; two years later they were making about 4s. 6d. (23 pence), an increase of 80 per cent.[10] In Midlothian wages varied from 2s. (10 pence) to 2s. 6d. (13 pence) a day, about 10s. (50 pence) a week in bad years to as high as 5s. (25 pence) a day, perhaps 30s. (£1.50) a week, in good.[11] In all the exporting coalfields wages dropped dramatically during the depression of the late 1870s. In Fife and Clackmannanshire they fell and fell until many men were making only 3s. (15 pence) a shift, even before any deductions were made for rent, coal, light and tools. It seemed obvious, conceded the *Colliery Guardian,* 'that for a man having a family to provide for, the problem of how to exist on this limited income, without falling into debt, is difficult of solution'.[12] Earnings were more stable in inland coalfields but even here there was a great difference between what a man could earn at the peak and in the trough of the trade cycle. In the boom of 1871-4 Black Country thick-coal miners were earning 5s. 6d. (27 pence) a day; some Midland men were said to be making 10s. (50 pence) a shift and 'Even public cadgers grumbled when they received coppers from colliers in Yorkshire.' But within just a few years the situation had changed completely. Thick-coal wages had fallen by a half and Nottinghamshire miners were taking home no more than 12s. (60 pence) a week and complaining 'of the wages, which are rendered so small by the half-time that they can scarcely support their families'. It was said in West Yorkshire that about one-third

> of the miners have been unemployed at most of the principal collieries. Reduction after reduction has taken place. Longer hours have been put upon the men, meal times have been knocked off, riddles [sieves used to separate the coal from the dross] have been taken into the pits, and all this with the idea of stimulating trade but which has yet proved futile.[13]

Uncertainty was the most important characteristic of miners' wages before the World War I. But almost as important, and just as much overlooked, was the wide range of wages paid within the industry. When a miner changed jobs at the pit, his earnings too were almost certain to change. Surface labourers were the lowest paid of all miners. They made perhaps three-fifths of a putter's wage and only about half what the hewer could expect to earn. In 1886 Lancashire surface labourers were earning 2s. 6d. (13 pence) a day as compared to the putters' 4s. 3d. (21 pence) and the hewers' 5s. (25 pence).[14] So in earnings as well as in status the surface labourer was at the very bottom of the colliery pile. Other surface workers were far better paid. Craftsmen could expect to earn about half as much again as the labourer — nearly as much as the hewer. At the Butterley collieries in the Erewash Valley in 1856, surface labourers were making 13s. – 14s. (65 – 70 pence) a week, hewers between 15s. (75 pence) and £1, enginemen from 15s. (75 pence) to 18s. (90 pence), banksmen just under £1, joiners and carpenters £1 and wheelwrights between 21s. and 22s. (£1.05 and £1.10).[15] And in addition to his cash pay, the mechanic, unlike the labourers, nearly always enjoyed both a free house (or a rent allowance) and free or cheap coal.[16]

Wage differentials were just as pronounced underground. When a boy first went below he started on a very low day-wage equal to a quarter or even less of a hewer's normal earnings in the same pit. Thus boys starting at the Bridgewater collieries north-west of Manchester began on just one eighth of a full day-wage.[17] The Children's Employment Commissioners found that throughout Ayrshire and Lanarkshire trappers were bringing home less than a quarter of the wages of even the youngest hewers and twenty years later twelve year old Midlothian lads going underground at the Dalkeith colliery were still receiving just a quarter of the full wage.[18] Yet it did not take the young miner long to start to earn better money. Within a year or two of going below he would probably start putting and within another couple of years would begin driving. Both putters and drivers were sometimes paid by the shift but more generally according to the number of tubs which they moved. In most parts of the country putters were employed by the hewers which makes it extremely difficult to discover how much they were paid. When working for an elder relative the putter probably received no more than a little pocket money.[19] When hired on the open market he was paid more and it was easier for a hard-working boy to play one hewer off against another.

> You went and worked with a man, with a collier getting the coal. The first man I worked with paid me half-a-crown a day. Then after a while, if you were any good, somebody else would say to you: "what's he paying you?" "Half-a-crown." "Right, come with me if you like, and I'll give you two-and-nine." ... Then if your mate is out, and the fireman sends you to somebody else, another collier, you'd say to this fellow: "the fireman sent me to work with you." "How much is he paying you?" Instead of two-and-nine, you'd say "three bob!" So you'd be trying to get a couple of coppers as you go through.[20]

In Northumberland and Durham the boys were paid directly by the colliery owner. At the Londonderry collieries in the 1840s one young driver was 'offered 5¾d. a score of 21 tubs'.[21] Later in the century a north-eastern authority explained that putters 'are usually paid from 11d. to 15d. per score of 6 tons, put an average distance of 80 yards with 1d. extra per score for every additional 20 yards'.[22] The good worker could still expect something extra. A Spennymoor man remembers that because he worked hard at driving, 'a putter came round to Seek me, on Saturday Morn to give me, an Old Time 3 Penny Bit, I thought I was in Clover.'[23] Although not easy, it is possible to make some estimate of putters' and drivers' pay, at least as a proportion of hewers' earnings. At the Glamorgan collieries for which information survives, oncost labourers earned 61 per cent of hewers' wages in 1869, 50 per cent in 1873 and 77 per cent in February—May 1878.[24] Then it has been calculated that in 1888 and 1914 the average wages of putters throughout the country were 3s. 7½d. (18 pence) and 6s. 5½d. (33 pence) a day, 76 and 73 per cent respectively, of the earnings of the piece-work coal-getters.[25] A rough generalisation suggests then that during the second half of the period at least, even the hardest working putter or driver would be lucky to earn much more than three-quarters the wage of a fully-fledged hewer.

The hewer of course was the aristocrat of the colliery labour force. If there was a colliery house available for example, it was he who got it. Hewers were paid in one of two ways. In the Black Country they were paid for the time that they agreed with the butty (the sub-contractor) that it would take them to cut a given volume of coal and in Scotland daily earnings were decided by multiplying the rate per ton by the estimated output, the 'darg', generally two tons in the east and three in the west.[26] Everywhere else the hewer was paid directly by the piece. Jevons explained in 1915:

The payment of wages in proportion to the amount of coal won or of work done is accomplished by means of a colliery price-list and by weighing the coal which each man hews, and "measuring-up" any other kind of work.

The workmen of a colliery who form themselves usually into a "lodge" or branch of the Miners' Federation of the coalfield, negotiate with the owner of the colliery a series of piece rates or "prices," to be paid for work done. The principal item is the "cutting price," or standard rate per ton paid for "getting" coal from the face in a properly opened stall or place. The "cutting price" may be anything from 1s. 1d. up to 3s. per ton, and is subject to the addition of a percentage. If the latter is 60 per cent. and the "cutting price" of a particular seam is 1s. 8d., and if the man sends up 2½ tons of coal in one day, he earns 6s. 8d. Thus:

$$
\begin{array}{lll}
 & \text{s.} & \text{d.} \\
\text{1s. 8d. x 2½} \quad = & 4 & 2 \\
\text{Add } \underline{60} \text{ x 4s. 2d.} = & \underline{2} & \underline{6} \\
\overline{100} & 6 & 8 \\
\end{array}
$$

The amount of coal worked by each man in charge of a stall is ascertained by weighing his trams at the pit head. He chalks his number on each tram when filled before it is taken away by the "putters" or "hauliers".[27]

Like the putters and drivers, but unlike those on day wages, the hewers were always trying to get paid for coal which they had not cut or moved. If paid by weight, they put stones into the bottom of the trams. If paid by measure, they roofed half-empty trams with lumps of coal.[28] In addition it was always a great temptation to try to find a flaw 'in the mode of keeping count of the tubs which the men filled and the putters brought out'. In mid-century Durham the count was kept by a system known as 'chalking' on.

The process was as follows: At every flat there was a boy or a man who was known as the "chalker on", whose duty it was to keep an account ... When a putter came out with a tub he shouted "Chalk on." The question was put, "Whe's the been at?" The reply was the name of the hewer. The initials of the hewer who had filled the tub were marked on the outside of the tub, ... [and on a board] There were three modes in operation in defrauding. The "chalker on", if he were so inclined, could mark more on than the man had filled or the putter had brought out, if he had a favourite, or "blood was thicker than water." Sometimes, when a putter came out and got the tub marked down to a man, he would quietly draw the tub back, and, resting awhile, would come up against the set with a bang, and shout out, "Chalk on," and another name. On a Saturday, if the "chalker on" wanted a short day, he would get to know by some process how many tubs each hewer wanted to fill, and in a short time the putter would be informed to tell the men they had the requisite number on the board, and they were enabled to get home early.[29]

The temptation was not removed by replacing 'chalking on' by a system of counting the hewers' and putters' tokens at the pit bank. Some men, like James Foster at the Raith colliery Cowdenbeath, exchanged other hewers' tokens for their own.[30] Others went a step further. It was discovered at Consett that tokens sent up between 4.30 and 5 p.m. were not counted until the following day. Very soon hewers were coming back to the pit in the evening to add their own tokens to those still to be counted.[31]

Methods of calculating wages and levels of earnings both varied widely between different parts of the workforce. Indeed wage differentials may well have increased somewhat in the late nineteenth century.[32] By 1892-4 the average weekly wage of all underground workers in South Wales was 23s. (£1.15). Hauliers however were bringing home no more than 15s. (75 pence) and time workers generally were often earning 30-40 per cent a shift less than the piece-rate hewers alongside them.[33]

Crucial as wage differentials between the different grades undoubtedly were, it would be a great mistake to assume that any two miners doing the same job in a particular pit would necessarily be earning even roughly the same. The earnings of hewers and all others on piecework naturally varied according to the skill, strength, age and inclinations of the individual. Newcomers always found it difficult to compete with experienced hewers. In 1905, for example, the large number of Polish immigrants in Lanarkshire were found to be earning well below the standard rate.[34] Earnings differed too 'because geological conditions varied so much from seam to seam, from face to face and (in the days of bord and pillar) from stall to stall'.[35] 'The place that is normal one week is abnormal next week,' explained a Somerset hewer. 'Faults, and other interruptions, cause the place to be abnormal, and the best skilled miner is at a disadvantage to produce his usual output.'[36] At the Marchioness of Londonderry's Old Durham colliery in 1863 there were 'places where the men are over their shoe tops in water, and nothing allowed for it'.[37] At the Fitzwilliam and Hemsworth collieries in South Yorkshire there were constant price-list disputes in the thinner seams but not in the easier and better-paid Shafton seam.[38]

Generally, allowance is promised so that the hewer may

earn his usual wage, as if his place was normal; but on pay days, after you have worked your hardest, and you are expecting a shilling or two above your day's wage, you find, to your dismay, that the allowance has been stopped.[39]

Another cause of variation in earning power was the sub-contract (or butty) system, whereby the butty agreed with the colliery-owner a price per ton for raising the coal and then sub-contracted out the actual work to the miners under him. With the development of deeper, more heavily capitalised mining, the system was abandoned in most parts of the country after about 1850. But it lingered on in the small, shallow pits of the Forest of Dean, Shropshire and the Black Country, where in 1908 a quarter of the entire labour force was still working under butties.[40] In these districts it was up to the butty to maximise his own profit by driving the miners as hard as he could. Playing one man off against another, he tried to pay as little as possible. At the very end of the century, for example, the contractor working one of the two pits belonging to the Florence colliery in North Staffordshire was still able to agree terms of payment with his men individually rather than collectively.[41]

Some wage differentials became institutionalised. Women, it is no surprise to learn, always earned less than men, even if they were doing exactly the same job. That self-styled champion of women's labour, A. J. Munby, found that in 1863 the average daily wage of Wigan pit brow lasses stood at between 1s. (5 pence) and 1s. 2d. (6 pence) and this at a time when men were getting from 2s. 6d. (13 pence) to 3s. (15 pence) a day for doing precisely the same work. Twenty years later it was the custom in Lancashire to pay surface women a penny a day for each year of their age, until they were eighteen years old. This meant that a female labourer could earn no more than 1s. 6d. (7 pence) a day compared to her male colleague's 2s. 6d. (13 pence). And when Munby's friend, Becky, started at the Woodhouse colliery, Wigan, she soon discovered that no woman was allowed to receive the free coal allowance which the men enjoyed.[42]

Yet another differential was that between married and

single men. Married men did much better with regard to the perquisites of free housing and concessionary coal. In the North-East, and in Scotland, married miners (or heads of households) received either free housing or a rent allowance. It has been estimated that by the end of the century this concession added between 10d. (4 pence) and 1s. 1d. (5 pence) per shift to a married man's earnings.[43] Married men all over the country were also favoured then it came to free coal. At Staveley, Derbyshire in 1858 any collier who was married or who had a housekeeper was given four hundredweights a week and in Shropshire it was the custom for every house-holder to be allowed one ton of coal a month.[44] By the beginning of this century this free coal was worth the equivalent of about 1½d. to 2d. (1 penny) a shift.[45] So the married miner in receipt of free coal and housing was getting a much greater reward for his labour than the single man. By the end of the period these two additional payments in kind were worth at least a shilling (5 pence) a shift, perhaps as much as five or six shillings (25 − 30 pence) a week.

If two men doing the same job in the same pit could be earning different wages, any move to another colliery would almost certainly mean a change in take-home pay. Physical conditions, labour supply, market opportunities, management techniques, union strength and profitability − all important determinants of miners' earnings − frequently varied very widely from pit to pit. New collieries often began by paying relatively high wages in order to attract labour while 'every explosion sends up the price of coal at that particular colliery'.[46] Nor did it take much to bring earnings down. As a pit grew older, it became necessary to work more difficult seams, it took longer to reach the face and it became more expensive to get the coal to the surface. And the men's wages fell accordingly. The marked wage differential that frequently existed between even adjacent pits can be seen clearly in South Yorkshire. Lundhill and Wombwell Main collieries were both working the Barnsley Bed within a mile or two of each other at Wombwell on the outskirts of Barnsley. Yet in 1879 Lundhill hewers were averaging 4s. 6d. (23 pence) a day while those at Wombwell Main were making 5s. 6d. (27 pence). There was a similar 25 per cent differential between

wages at Houghton Main and Carlton Main, two more collieries which had been sunk to the Barnsley Bed about six
miles apart in 1872-3. At the one pit in 1900, hewers were
averaging 7s. 4d. (37 pence) a day; at the other they were
able to make 9s. 3d. (46 pence).[47] It is true that as the century
progressed earnings at different collieries did tend to assume
a more common pattern.[48] But differences persisted and it
was always worth the miner's while to consider moving.
Even in 1914 day and piece-rates were generally fixed, not at
district level, but at the individual pit as the result of negotiations between the members of the union lodge and the
manager of the colliery.[49]

Still another important wage differential existed in the
first half of the period. Miners in several parts of the country
were either paid in kind or, if paid in cash, were compelled
to spend a proportion of their wages in the employer's
'tommy' shop. This 'truck' system was most common at the
small Black Country pits and in the new mining communities
which sprang up in Durham, South Wales and in the West of
Scotland, where in the 1860s almost half the workforce was
subject to truck.[50] In these coalfields many miners were
encouraged, or even forced, to shop at the employer's store.
The men and boys at James Merry's colliery at Ardeen in
Ayrshire were paid monthly in cash. But if they did not
shop at the colliery store they soon found that they were
refused any advance between pays.[51] Elsewhere in Scotland
the pressure was more direct; those declining to spend a certain part of their earnings in the tommy shop were warned
and then dismissed.[52] In the Aberdare district of South Wales,
too, men were compelled to buy goods from the company
store. If they needed cash then they had to resell these goods,
perhaps at as much as a 50 per cent discount.[53] An orphan
from Blaina, Janet Jones, caused a sensation in 1864 when she
told a Tredegar court of the the two years during which her
job had been to move coal tubs around the pit brow,

> My wages have been 5s. 6d. a week. During the whole
> time, I did not receive any money. I never received a
> farthing in money, but all the goods at the shop. The
> whole of my wages were swallowed up in shop bills for
> bread and tea, on which I lived.[54]

Monopolistic and semi-monopolistic company shops were able to sell poor quality goods at inflated prices with minimal standards of service, or even civility. In South Wales for instance,

> When the workman was not in credit and visited the shop in the hope of getting a quarter of tobacco . . . he would get a rough time. He would be met with a stern refusal followed up by an invitation to "go to hell". This was the only kind of free invitation these poor fellows ever got.[55]

Truck shop prices were never low and rarely even competitive. Thus in the Black Country some owners took advantage of the depression of 1816 to pay wages in truck and to charge 15 to 20 per cent above the market price for provisions.[56] A few years later it was alleged that local tommy shops were selling 6s. (30 pence) bags of flour for 9s. (45 pence) and charging 8d. (4 pence) a pound for 'very poor' bacon when other shops were charging only 6d. (3 pence) a pound for best bacon.[57] And at a Sheffield meeting in 1850 the Yorkshire miners' leader, David Swallow, banged a railing with a piece of board-like meat which he dubbed 'tommy shop' bacon.[58]

Truck began to break down around the middle of the nineteenth century as improved communications made pit villages isolated, as copper and silver coins became more plentiful, as anti-truck legislation became more effective and as miners formed their own attractive alternative to the tommy shop, the co-op, a development which will be analysed in some detail in the next chapter.[59] But this must not be allowed to diminish the importance of truck. Throughout the first half of the century in South Wales, Durham and the West of Scotland, and far later in the Black Country, the truck system seriously diminished the spending power of tens of thousands of miners and their families.

It will be clear by now how difficult it is to make even the most limited generalisations about miners' earnings during the nineteenth century. They varied from year to year and from week to week, from pit to pit, seam to seam, stall to stall and from man to man. It is even more difficult to compare earnings between coalfields for there were important regional variations in, for example, the prevalence of truck and the provision of free housing, variations which directly affected the level of the miners' disposable income. Yet the regional perspective is essential. It is necessary to fight a way through the morass of impressionistic, confusing — and often conflicting — evidence of earnings in order to understand the domestic and industrial experiences of miners in the different coalfields.[60]

The miners of Northumberland and Durham were the best paid in the country. Their money wages were (comparatively) high and nearly every married man received a free colliery-owned house or, after the labour influx of the early 1870s, a rent allowance in lieu. The workmen benefited, too, from the North-Eastern owners' unique system of relief to the many underground workmen injured during the course of their employment. The injured miner was allowed to continue to live in his colliery house and to receive his allowance of free coal. And for a substantial period, generally a year, he was paid an accident allowance, known as smart money, of about 5s. (25 pence) a week (half this sum for boys). The payment of smart money alone represented a substantial addition to the employers' wage bill. In the year 1879-1880, for example, 7,942 injured men were paid well over £6,000, a sum equivalent to almost 7s. (35 pence) for every man, and almost 11s. (55 pence) for every boy, among the North-East's 94,000 strong labour force.[61]

The depradations of the truck system in early nineteenth-century Durham were largely offset, at least in good years, by the owners' practice of providing free drink and paying a bounty at the time of the annual binding. Whatever the other drawbacks of the bond, the payment of binding money was always welcome. In 1800 the usual sum throughout the coalfield stood at between one and three guineas, (£1.5 and £3.15), equivalent to between 4¾d. (2 pence) 1s. 2¼d. (6 pence) a week for the whole year. Five years later maximum payments were fixed. Married hewers were to be paid no more than three guineas (£3.15) on the Tyne and five guineas (£5.25) on the Wear; single hewers could be paid up to £3. 13s. 6d. (£3.68) on the Tyne and up to six guineas (£6.30) on the Wear; oncost and surface workers being paid proportionately less. In good years, however, the competition for labour soon forced up the level of binding money. In both 1809 and 1810 hewers in the Newcastle district received five guineas (£5.25) each, a sum representing very nearly an extra two shillings (10 pence) a week throughout the whole of the year.[62] So despite the truck system and the often violent fluctuations of the export trade, the North-East pitman, with his high money wages, his binding money, his colliery house and his smart money, was far better rewarded than his fellow miners in other parts of the country. Later chapters will explain the crucial effect which this comparative affluence was to have on the domestic arrangements and the trade union organisation of the Northumberland and Durham miners.

Money wages were nearly as good in the rapidly expanding Midland and South Wales coalfields. They were especially high in new areas like the Rhondda: in 1871 for example the Rhondda owners were offering wages 10 per cent above those paid in the Aberdare Valley and a full 25 per cent higher than those being offered for similar work in the Merthyr Valley. But in both South Wales and the Midlands truck was common in the first half of the century. In both districts, too, it was unusual for the owners to provide accommodation and almost unheard of for them to make any provision for their injured workmen. In real terms then, few Midland or South Wales miners could hope to approach the level of earnings enjoyed by the North-Eastern pitmen.

Other relatively well paid miners were to be found in Yorkshire, Scotland (specially Lanarkshire) and, contrary to popular belief, in Lancashire. The level of real earnings in Scotland was considerably increased by the early nineteenth-century practice of paying binding money and by the widespread adoption of the colliery house system. Between a third and a half of Lanarkshire miners lived in free or cheap accommodation, a figure which rose to as high as three-quarters in Ayrshire and the Lothians. Earnings were lower in the Forest of Dean and the West Midlands. There were few colliery houses; there was little relief for injured miners and truck remained common in the Black Country during the whole century. (Equally persistent though was the employers'

custom of giving their men a free daily ration of four pints
of beer. Indeed when nationalisation finally came in 1947
the National Coal Board inherited contractual obligations
towards some South Staffordshire men in lieu of their previous allowance of beer.[63]) Wages were particularly depressed
in Somerset and North Wales. From the 1880s, if not before,
Somerset hewers were making only about as much as oncost
workers in Durham or South Wales and even less than the
oncost men employed in Northumberland, Lancashire, or
South Staffordshire. Earnings in low-wage districts like
Somerset and North Wales were at best little more than three-
quarters, and at worst scarcely half, those being paid for
comparable jobs in high-wage coalfields like the Midlands,
South Wales and the North-East. This wage diversity, it will
be seen, is crucial to any understanding of the distinctive
social and labour histories of the various coalfields.

Whatever the regional variations in miners' earnings, there
is no doubt that miners generally were well paid compared
to most other workers. Mining wages were often half as much
again as those paid to agricultural labourers. At the end of
the nineteenth century, farmworkers in Nottinghamshire's
Leen Valley were making around £1 a week whereas local
face-workers were able to earn nearly 8s. (40 pence) a shift.[64]
It was only during the very worst depressions in the coal
industry that miners were ever prepared to compare their
wages unfavourably — and then probably only rhetorically
— with those of agricultural labourers.[65] Miners were better paid, too, than general labourers and factory workers.
Labourers employed on the Thornsett (Derbyshire) Turn-
pike in 1832 were earning 2s. (10 pence) a day, compared
to the local miners' 3-4s. (15-20 pence). Even in the low-
wage North Wales coalfield miners could hold their own
against other local workmen. In 1850 Flintshire fillers [who
filled coal at the face] were making 15s. (75 pence) a week,
2-3s. (10-15 pence) more than unskilled brickmakers.[66] The
8s. a week which Lancashire drawers were said to be making
in 1841 was not very much. But it was more than labourers
were able to earn in the nearby factories.[67] Lancashire miners
knew their own worth.

Collier lads get gold and silver.
Factory lads get nobbut brass.
Who'd get married to a two-loom weaver
When there's plenty of collier lads?[68]

There is no doubt either that the disposable income of miners in every coalfield grew substantially during the course of the nineteenth century. Money wages rose as truck declined and fining became less severe. The pitmen of Northumberland and Durham were the largest and best paid body of miners in the country. The most reliable evidence suggests that in 1795 they were making from 2s. 6d. to 3s. (13-17 pence) a day, 18s. (90 pence) a week at best. By 1888 even oncost workers were making more than this and hewers were bringing home about 26s. (£1.30) a week. Thereafter earnings rose more rapidly and at the end of the period oncost workers were averaging 6s. (30 pence) a shift and hewers as much as 9s. (45 pence), perhaps over £2 a week. In 1914, then, the North-Eastern pitman was probably earning three times as much as his predecessor at the end of the eighteenth century.[69] An examination of earnings in the rapidly-expanding South Wales coalfield reveals a similar trend. Hewers in 1800 were able to earn about 2s. to 2s. 6d. (10-13 pence) a day, 15s. (75 pence) a week at the very most. This was said to have increased to just under £1 a week (perhaps 3s. — 15 pence — a day) in 1840, to 4s. 10d. (24 pence) a day in 1888 and to as much as 9s. 4d. (47 pence) a shift at the beginning of World War I.[70] So it is clear that in South Wales, too, earnings increased three-fold during the nineteenth century. The rate of increase was nearly as great even in small, slowly expanding coalfields like Cumberland. In 1800, hewers in that area were probably making about 3s. (15 pence) a shift, between 12s. and 15s. (60 pence and 75 pence) a week. By 1888, they were earning 4s. 5d. (22 pence) and by 1914, 8s. 2d. (41 pence), up to about £2 per week.[71] Nor is there any doubt that over the country as a whole the increase in miners' earnings during the course of the nineteenth century was more — far more — than sufficient to keep pace with the cost of living.[72] Between the 1790s and the 1840s, it is true, the increase in money wages was probably only just enough to maintain the level of real wages. But thereafter the cost of the living tended to fall until the middle of the 1890s, a period during which earnings were climbing fairly sharply. Food, which always loomed large in the family budget, was becoming appreciably cheaper. The Co-operative Wholesale Society estimated that the average family grocery order, which would have cost 7s. 6½d. (38 pence) in 1882, cost only 5s. 11d. (30 pence) in 1888 and as little as 4s.-10½d. (25 pence) in 1895. But the cost of living began to increase again from the mid-1890s, with the co-op's typical order rising to 6s. 4¼d. (32 pence) in 1914. It is almost impossible to believe in our times of constant inflation but prices were back roughly to where they had been a century or more before.

This discussion of changes in earnings during the nineteenth century has been based of course on averages, averages which by their very nature tend to smooth out anomalies, fluctuations, difficulties and differences. We have warned how misleading it is to generalise even about the earnings of one or two men doing the same job in the same pit. Yet if we ignore our own warning and try to produce one generalisation that will cover every workman in every coalfield throughout the whole of the nineteenth century, it must surely be that the underlying trend of earnings and of real wages was ever upwards. That mythical beast, the 'typical' British miner, was earning three times as much, and was three times as well off, in 1914 as he had been in 1800.

4

The Mining Settlement

There is a tendency to think of the traditional nineteenth-century mining community as being synonymous with all that is dreary and depressing. We think of a remote village or small town, built on a grid-iron pattern, with endless rows of insanitary, colliery-owned terrace houses, a few tawdry shops, lots of pubs, a chapel and perhaps a school, all dependent on one, or at most a handful of nearby pits. We think of places like Maerdy, Treharris and Senghenydd in Glamorgan; Hoyland, Wombwell, Dodworth and Denaby Main in South Yorkshire; New Hartley, New Seaton and Seaton Delaval in Northumberland; and Brancepeth, Hetton-le-Hole, Wingate and Thornley in County Durham.

Hetton-le-Hole, two miles south of Houghton-le-Spring and six miles north-east of Durham, may stand as typical of the stereotype.[1] Hetton virtually owed its existence to the Hetton Coal Company which was formed in 1819 and immediately started sinking operations through the magnesian limestone. Almost at once Hetton, with one pub for every thirty or so inhabitants, began to display some of the characteristic features of the traditional mining settlement. The Lyons colliery was opened in 1822 (to be followed in 1833 by the Eppleton and later by the Elemore) and within a decade the company became one of the leaders of the North-East coal trade. The population of Hetton grew dramatically, from 919 in 1821, to 5,887 in 1831 and 12,726 in 1891. Long, dreary rows of grey, stone and brick cottages were thrown up. By the 1870s the company owned about 350 houses, 200 of which clustered in rows around the centre of the village. Most of the streets were unpaved and undrained and the majority of families were forced to share both

81

toilets and water taps. When Sir Arthur Redmayne recalled the eight and a half years he spent in Hetton during the 1880s he remembered:

> the discomfort and inconvenience of the dwellings provided for the miners in those days — a lack of adequate water-supply, one outside tap shared between several householders, communal ash-bins close to the cottage doors, bad roads and a dreary outlook — for pit-heaps dotted here and there, with their monotonous regularity of outline and colour, cannot be said to have other than a depressing effect on the landscape. There was also an entire absence of buildings of any comeliness — the church of drab yellow brick, or an edifice of wood and corrugated iron, the Methodist Chapel, perhaps, a little more impressive, though also frequently of the same drab colour. Few villages then possessed a Mechanics' Institute or Workmen's Club.[2]

Hetton certainly seems to conform to the traditional stereotype of the nineteenth-century mining village or small town. But how accurate is the stereotype? Did the miners really live in the small, isolated, featureless, single industry wilderness that survives in the popular imagination? The reality is rather more complex. Nineteenth-century mining settlements varied enormously in their design, their degree of isolation, the extent of their dependence on a single economic activity, the quality of their sanitation and housing provision and the range of their other amenities. The 'typical' mining settlement was common, never mind typical, only in certain parts of the country — in prosperous, expanding, relatively high-wage districts like the North-East, Lanarkshire, South Yorkshire, South Wales and parts of the Midlands.

It has become a commonplace to comment on the geographical isolation of places like Hetton. 'Mining settlements were often communities apart . . . many were sunk in remote areas. This isolation shaped both personal attitudes and community development to make the miner . . . into a conservative.'[3] Certainly many miners in Northumberland, Durham, Scotland, Lancashire, South Yorkshire and South Wales did have their homes in separate villages or small towns.[4] But it is only too easy to exaggerate the isolation of these colliery communities. Hetton itself was only two miles from the much larger Houghton-le-Spring and few, if any, colliery villages were ever more than a few miles from a non-mining settlement. From the middle of the century the railways broke down still more the isolation of all but the most remote villages. 'We have no mine in Scotland', admitted Alexander McDonald in 1863, 'that cannot be reached from town in half-an-hour.'[5] In many parts of the country it was unusual for miners to live in separate communities. The early part of the century saw many Midland mining settlements grafted onto existing villages and small industrial towns.[6] In the Black Country, in the Rhondda, along the valleys of the Dearne and Don in South Yorkshire, and in parts of Lanarkshire and North Staffordshire the miners were lost in a vast tide of sprawling urbanisation.[7] Some colliery villages became absorbed by a larger neighbour. The Durham villages of Preston and Percy Main were swallowed up by North Shields while Spital Tongues colliery, which was sunk on the outskirts of Newcastle in the 1840s, had completely lost its separate identity thirty years later.[8] And it must not be forgotten either that miners all over the country lived in large towns and cities. Many Yorkshire men for example commuted to work from Bradford, Barnsley, Leeds, Wakefield and Castleford. For years more than 500 men travelled every morning from Barnsley out to the Carlton and Monk Bretton collieries.[9] By 1911 over a third of all Lancashire mineworkers had their homes in the large towns and cities of Manchester, Burnley, Oldham, Bolton, Blackburn, Wigan and St Helens.[10]

To claim then that nineteenth-century miners lived exclusively, or even predominantly, in isolated pit villages is seriously misleading. Some lived in tiny rows in the middle of the countryside while others lived in the middle of huge cities. Nearly all, however, were within relatively easy reach of the amenities and facilities of at least one sizeable town. This, in turn, immediately throws into question many of the assumptions commonly made about the dependence of mining communities on a single economic activity or even upon a single employer.[11] Even in those districts associated most closely with the coal industry, the miners were easily outnumbered by other workers throughout the first half of the century. In this period the highest proportion of miners was to be found in Durham, Glamorgan and Monmouthshire; yet even here five out of every six men had nothing whatsoever to do with the coal industry. Even at the very heart of these coalfields, at colliery towns like Merthyr and Houghton-le-Spring, the proportion of miners to other workers never reached as high as one in three.[12] It was only in the last years of the century that mineworkers ever became the largest occupational group in a particular district, and then only rarely. But in the few areas where they did predominate, their influence could be overwhelming. In 1911, for example, mining accounted for more than 70 per cent of the male working population of Northumberland and Durham communities like Ashington, Bedlington, Earsdon, Tanfield and Ryton; and of small South Yorkshire towns like Wombwell, Darfield, Darton, Royston and Bolton-on-Dearne.[13]

> [In] the Rhondda Valley . . . with a population . . . of more than 150,000, we find that about 95 per cent. belong to families engaged in, or dependent upon, the mining industry. There are few works or factories, or other employment except for the comparatively few openings for employment on the railways and in shops.[14]

These, however, were the exceptions. It remained uncommon for mining communities to be totally dependent on the coal industry and even more unusual for them to be controlled by a single employer. This is not to deny of course that colliery owners could — and often did — wield very great economic, political and social power. Nor did this power decline as the century progressed. Northumberland and Durham pit villages were often dominated by one employer and 'company towns' grew up elsewhere as the industry expanded into new areas. The Glamorgan Coal Company built a model community at Llwynypia in the Rhondda.[15] During the opening up of the concealed South Yorkshire and Nottinghamshire-Derbyshire coalfields, the Bolsover Company built Creswell near Welbeck, the Brodsworth Main

Company ran the Woodlands model village near Doncaster and the Denaby Main Company founded the South Yorkshire village of Denaby.[16] In 1885, the local lodge secretary claimed, 'The Denaby Colliery Company own the whole of the property in the village with the exception of two houses.'[17]

It is not generally appreciated that no matter in which part of the country he lived, the miner nearly always had access to pits run by a number of different owners, even if towards the end of the century this might mean a train journey to and from work. Between 1871 and 1877 fourteen new collieries (with thirty-two pits) were opened within a two miles radius of Hamilton West railway station in Lanarkshire.[18] By 1914 there were eight collieries within half an hour's walk of Barnsley town centre (see Map 8).[19] At nearby Hoyland Nether three of the six local collieries were owned by Earl Fitzwilliam, but the other three belonged to the Newton Chambers Company, the Hoyland Coal Company and the Wharncliffe Silkstone Coal Company.[20] There was even some degree of choice in Hetton. A miner brought up in the village recalls that by the end of the period there were two firms other than the Hetton Coal Company running pits 'within a reachable distance' of his home.[21]

It is clear therefore that the widely-held view that most nineteenth-century mining villages were dependent on a single powerful coal-owner stands in need of some modification. And it is reasonable to go on to question the equally common belief that 'The typical mining village was a dreary collection of box-like cottages, arranged in monotonous rows, each identical with the next.'[22] It is true that the stereotyped grid-iron pattern was common in Northumberland and Durham and wherever else it had been found necessary to house a large body of workers quickly and cheaply in rural areas. The pit rows bit in to the countryside of every major coalfield.[23] A Durham miner remembers the 'row after row of colliery houses' at Wingate.[24] The Medical Officer of Health for Ayrshire deplored the county's mining villages which, he said, were 'frequently arranged in several parallel rows all facing the same direction; others may face each other on opposite sides of the road; or they may be found in single

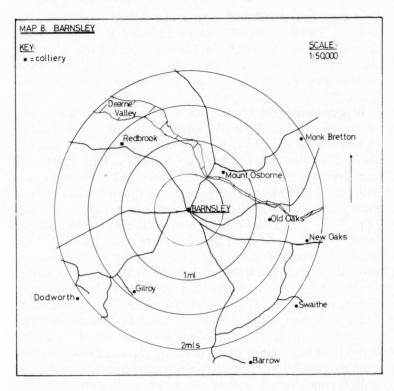

MAP 8. BARNSLEY

KEY:
• = colliery

SCALE:
1:50,000

Dearne Valley

Redbrook

Monk Bretton

Mount Osborne

BARNSLEY

Old Oaks

New Oaks

1ml

Gilroy

Dodworth

Swaithe

2mls

Barrow

rows or squares.'[25] Yet this unimaginative grid-iron design was never universal. Even in purpose-built mining villages in Northumberland, the very home of the grid-iron, other lay-outs intruded. Walbottle was arranged round a village green with a duck pond in the middle. The centre of Bedlington contained a conglomeration of old and new property, much of it tenemented.[26] In other places in the North-East rows were built in haphazard fashion, at odd angles to each other; while any one row might contain houses of different ages, different sizes, different materials, all in different states of repair.[27] Everywhere the grid-iron had to give way before the activity of speculative builders and the enthusiasm of home-improvers. The same miner who recalled the 'row after row of colliery houses' at Wingate remembered that his father 'had a lean-to built adjoining the pantry which eventually became a wash-house-cum-storeroom'.[28]

Some owners, too, showed an awareness of the need to break away from the grid-iron. At Gartsherrie in Lanarkshire Messrs Baird built 250 houses in two large squares, the space within each square being laid out as gardens, with a further large plot left available for cricket and other recreational purposes.[29] When later in the century the Bolsover Company built Creswell on the borders of Derbyshire and Nottinghamshire, there were

> 280 two-storied cottages, built in the form of a double octagon, there being an inner and outer "circle". . . The houses vary in design . . . Within the inner "circle" is a large green, which is relieved by shrubberies at intervals. A pretty, artistically designed band-stand occupies a position in the centre . . . There is also a circular playground in which the children can disport themselves.[30]

At Woodlands near Doncaster the Brodsworth Main Colliery Company

> acquired a large area of land near the collieries, laid it out on Garden City lines with pleasant streets, lined with trees and grass plots, and with open spaces and recreation grounds. The cottages, of which there are . . . over 630, are built in pairs, and in small groups of four or five; and the average number per acre is less than six, which is to be compared with 25 to 30 to the acre by the present-day [1915] speculative builder.[31]

A glance at any ordnance survey map of the coalfields will reveal that these well-known model mining villages were really little more than a minor variation on the traditional high density lay-out. But they do suggest a growing awareness of the possibly corrosive effect of living in one of rows upon rows of almost identical houses. They are important, too, because they throw into relief the fact that the grid-iron pattern, though common, was never universal.

It is possible to see now that the usual view of nineteenth-century mining communities stands in need of serious revision. Some miners lived in single rows in the middle of the country.[32] Many more lived amid the urban sprawl or in bustling towns and cities. The isolated, purpose-built, grid-iron village or small town dependent on coal — or even on a single colliery — was built, it must be admitted, in every major coalfield. But this 'typical' nineteenth-century mining settlement predominated only in parts of Northumberland, Durham, South Yorkshire and South Wales.

Equally misleading is the view which suggests that nineteenth-century mining communities were bereft of every amenity save one, the iniquitous public house. Certainly it is true that whenever a colliery settlement was established, there was a time-lag before the provision of social, commercial and spiritual amenities caught up with demand.[33] There was at first always a serious shortage of shops, churches and other public buildings. But not for long. Thrown back on to their own devices, the miners worked hard, and increasingly successfully, to improve the quality of life in the villages. Chapel building is a good example.[34] The mine owner or other wealthy patron might provide a site and some of the building materials.[35] But the major cost of building was borne by the miners and their families. The bulk of the burden — both physical and financial — fell upon the ordinary chapel member. The mines' commissioner noted that in South Wales 'one neighbourhood assists another in building these chapels, and . . . the funds to build and support them come almost exclusively from the savings of daily labour.'[36] In the early 1840s the men at the Redding colliery in Stirlingshire were subscribing towards the cost of a methodist chapel.[37] More often it was the women who did the fund-raising. In one small north Durham village the miners' wives collected money for a new Primitive Methodist chapel by holding sewing parties in each others' homes and selling the goods they made at chapel bazaars.[38] Everywhere the mining community sought to remedy the weak parochial structure of the Anglican church. Soon even 'typical' mining villages like Hetton-le-Hole were able to boast 'chapels for Wesleyans, Primitive Methodists and Wesleyan New Connexion.'[39] By the middle of the century almost every colliery community in the country had the use of at least one chapel built and paid for largely by its own members.

More striking still was the contribution which the miners made towards the improvement of shopping facilities. There was a good deal to do. Owners in Lanarkshire, South Wales, Durham and the Black Country, who had tasted the easy profits of tommy shopkeeping, were often reluctant to make way for more normal trade. In 1850 the Tredegar, Rhymney and Ebbw Vale Companies still had clauses in their leases preventing houses on their land from being used for any sort of business without their consent.[40] But whenever free competition was allowed in a settlement of any size, a good number of shops would soon be opened. In this the miners and their wives themselves played an important part. It is well known that blacklisted union leaders often went in for shopkeeping, at least in the short-term. The famous Durham trade unionist, John Wilson, was refused work in the 1870s so together with his wife and two daughters he took over a shop ('largely stationery . . . and a sort of general dealing') at Wheatley Hill, between Durham and Hartlepool.[41]

Far less is know about the many ordinary families who saw in shopkeeping the opportunity to make a little extra money on the side. Oral testimony and autobiographical evidence do however reveal a little about this almost completely ignored facet of working-class self-help. A South Wales hewer recalls in his autobiography that before World War I, he 'saw only a very few real shops in the village — most of the selling was done in houses that had converted their front rooms into shops and hung outside signs advertising tea or salmon.' Then just after the war,

> my wife had caught the craze for shop-keeping, so I had made a counter and shelves, and our front room was a shop — it was well stocked too. We had reared two bacon pigs, and we were selling the bacon, and had peas and poultry out in the garden.[42]

This was common practice. At the beginning of this century a good proportion of all the shops and services in Essington — a small, undistinguished mining village on the outskirts of Wolverhampton — were provided by miners or their wives. The butchers and the post office were run by professionals but several of the grocers were carried on by colliers' wives and the village barber could only open in the evening when he had finished his work at the pit.[43]

Anybody with a marketable skill tried to turn it to good account.

In a [Durham] mining village it was common to find the lamp-repairer of the colliery, doing tinker repairs at home. Workshop in yard or garden. Repairs such as soldering bottles, kettles, tin-baths, pails etc. — The joiner mending broken chairs, tables etc. also sharpening saws, chisels, knives and scissors.[44]

In villages near the coast 'people tried to make a little bit on the side to help out in those days. Some would make toffee, some would make ginger beer, others would boil and sell crabs and winkles, many sold fish from their cobbles.'[45]

Fortunately the historian is not dependent solely on the memories of elderly miners and their families. Just occasionally, more conventional sources throw up evidence of unusual variants of what was clearly a large scale, if generally unspectacular, form of entrepreneurial enterprise. At West Cramlington in Northumberland one miner converted his kitchen garden to house a poultry farm and a cold shower which he hired out as a public bath during the prosperous 1870s.[46] A few years earlier the *Colliery Guardian* had reported an even more unusual undertaking in South Yorkshire. At Elsecar,

a very intelligent working collier has fitted up a Turkish bath, in a very effective manner, and which is extensively patronised, nearly 200 colliers having enjoyed the Eastern luxury during the three weeks it has been in operation.[47]

Keeping shop could be a chancy business though. Although bad debts were common, it was difficult to refuse credit to ex-workmates. Evidence survives of at least one Yorkshire miner who learnt this lesson the hard way. A hewer went into business as a grocer at the mining village of Ryton, only to go bankrupt when the coal industry fell into one of its periodic declines and many of his customers left the village without paying their bills.[48]

To many mining families, opening a small shop must have seemed perhaps the only way of escaping or at least easing their lot in life. Yet it was not only by becoming shopkeepers themselves that the miners helped to improve the quality of shopping in the colliery villages and towns. They did a great deal as well by their support of the co-operative movement.

At first there were formidable obstacles to face. Not only did many colliery-house leases in the 1850s and 1860s prohibit the opening of shops but the owners and their agents were often openly hostile. At Seaton Delaval in Northumberland a clause in the royalty stipulated that only one shop could be built on the colliery's land with the result that when the co-operative store did finally open in 1863, it had to be built at Seaton Terrace, nearly a mile away.[49] The owner of the Brancepeth colliery near Durham was connected with shops dealing in meat, groceries, painting and glazing and boot and shoe-making. So no competition was welcome and as soon as he learned that some of his men were planning to start a co-op in the neighbourhood he decided to give them the sack.[50] The miners persisted however and during the boom of the early 1870s some North-Eastern employers began to offer the nascent movement their tentative support. The Boldon colliery co-op ran its store in an end-house which it received rent-free from the owners.[51] Meanwhile the Birtley Iron and Coal Company helped in the formulation of the Pelaw Main Co-operative Butcher Society by donating free premises, a horse and a small amount of money.[52]

The co-op offered solid advantages over its rivals. Although there was no credit, the goods on sale were always genuine, there was no underweighing, prices were competitive and there was always the dividend to look forward to.[53] This was the great thing. At the Hindley Friendly Co-operative Society store outside Wigan in the 1860s all shoppers, whether or not they were members, received a dividend of 1s. 4d. (7 pence) in the pound on purchases of groceries, meat, drapery, shoes and clogs.[54] So there was always something to fall back on. At the beginning of this century the Walsall Co-op's dividend stood at about 2s. 9d. (14 pence) in the pound and in bad times children were sent from the South Staffordshire colliery villages to the nearest branch to collect whatever was due to their parents.[55]

In 1850 there were more than 200 retail co-operative societies, mainly in the north of England, and by the end of the century the movement had well over a million and a half members.[56] The coming of the co-op often marked a village's arrival as an established community. The 1850s and

1860s saw co-ops opening round Barnsley and later, as pits were sunk across the river Dearne, so stores followed at Royston (1888), Cudworth and Hemsworth (1892), Carlton (by 1897), Ryhill (1899) and Great Houghton (1900).[57] But it was in the relatively high-wage and relatively isolated mining communities of Northumberland and Durham that the co-operative movement made its greatest impact. For instance the Cowpen District Co-operative Provident Society was started in 1861 when a dozen men paid a shilling each to open a shop in the pantry of one of their homes. Within eighteen months the society claimed 140 members, had converted a six-room house for use as a store and felt confident enough to predict both a dividend of 2s. (10 pence) in the pound and a payment of five per cent on all paid up shares.[58] By 1863 there were co-ops at many other North-Eastern colliery villages; at Bedlington, Cramlington and Dunston in Northumberland for example, and at Blaydon, Hetton, Low Fell, Tanfield and Trimdon Grange in Durham.[59]

The coming of the co-op had far more impact than a mere catalogue of names and dates might perhaps suggest. In certain parts of the country, specially in the North-East, membership of the co-op came to be extremely common. In the boom year of 1873, the 470 members of the Hetton co-op represented one in seventeen of the village's total population; the 452 members of the Bedlington co-op one in fourteen of its population; and the 1,800 or so members of the five branches of the famous Cramlington co-op almost half the people in that village.[60] Indeed in some of the model colliery villages built towards the end of the century membership seems virtually to have constituted a condition of employment: every one of the 1,400 men and boys employed at Creswell in 1900 was said to belong to the local co-op.[61] Whether co-op members or not, miners and their families living in the 'typical' mining settlements of South Wales, South Yorkshire, Lanarkshire and the North-East all stood to gain from the increased competition which the new form of shopping brought with it. The confident 1863 prediction of the leaders of the Killoe (Durham) Industrial Co-operative Society was to be fulfilled: 'We will live better and wear better, and at a great deal cheaper rate, than we can do under the village grocer, with his few bullets for the child, a few pins for the mistress, or a few pipes for the old man.'[62] Every miner and his family benefited too from the example of what they themselves could do to improve the facilities of the colliery villages. Jack Lawson remembers going to listen to 'the quarterly meeting discussions' of his local co-op, 'little thinking I was merely learning the great art of self-government'.[63] The co-operative store, 'often the largest and most impressive building in the district', came to stand with the chapel, and later with the institute and the union hall, as a symbol of the miners' growing interest in both the principle and the practice of self-help.[64]

The miners' involvement in self-improvement was also shown in their efforts to purchase their own homes. While it is well-known that home ownership was comparatively common among the miners of South Wales, it is generally assumed to have been almost non-existent in other coalfields. Most miners, it is believed, lived rent-free — or very cheaply — in houses belonging to their employers. Like most generalisations made about nineteenth-century miners, this one is no more than partly true. Many miners in Scotland and the North-East of England did live in colliery houses, the proportion varying from a third to a half in Lanarkshire and central Scotland to as high as three-quarters in Ayrshire, the Lothians and Northumberland and Durham.[65] In other coalfields it was unusual for miners to live in a colliery-owned house for in most places the employers provided houses for only two purposes; to attract labour when a pit was first opened, and later to retain certain key groups of workers.[66] Colliery houses were particularly scarce in West Yorkshire and, despite the continual shortage of labour, in the rapidly expanding South Wales coalfield.[67]

The standard of colliery-owned housing varied widely. It was at its best on the estates of aristocratic coal-owners and in the model village built towards the end of the century. In 1811 the Earl of Moira constructed two terraces known as Stone Row at Moira in Leicestershire. Each of the thirty-eight houses had two bedrooms upstairs with a parlour, a kitchen, a large front room and a coal cupboard downstairs. The stone external walls were a good eighteen inches thick and outside each house there was an earth closet, an ashpit and a long garden.[68] Earl Fitzwilliam provided his South Yorkshire miners with homes . . .

> . . . of a class superior in size and arrangement, and in the conveniences attached, to those belonging to the working classes. Those at Elsecar consist of four rooms and a pantry, a small back court, ash-pit, a pig-sty, and a garden; the small space before the front door is walled round, . . . a low gate preventing the children from straying into the road. Proper conveniences are attached to every six or seven houses . . . The gardens, of 500 yards of ground each, are cultivated with much care.[69]

At Dalkeith colliery in the Lothians the Dowager Lady Ruthven built model homes fifty-three feet by twenty feet, each containing a kitchen, living room, a small scullery and a separate water closet.[70] Those owners with an awareness of the need to break away from the stereotyped grid-iron pattern often also showed a desire to offer reasonable sanitary facilities. Messrs Baird provided gardens, laid on an ample water supply and employed men every day to clear the refuse from their 250 houses at Gartsherrie.[71] The Brodsworth Main Colliery Company's model cottages at Woodlands each had three bedrooms, a bath with hot and cold water and good-sized gardens to both front and rear.

> The site is a well-timbered estate of 120 acres, mostly park land with fine old trees, and includes a mansion house and a lake. On it, well distributed over the whole area, there have been built 964 houses, giving an average of about eight houses to the acre. Ample space prevails everywhere . . .
> The houses are built semi-detached or in blocks of three, four, or five houses. There are more than twenty different types of house presenting a considerable variety of tasteful design. They may justly claim to be architecture, as distinguished from mere building.[72]

These, however, were the exceptions. Most colliery houses could not 'claim to be architecture'. And the tenants of many model colliery houses were subjected to a very high level of paternalistic interference and regulation. The Dowager Lady Ruthven manipulated a combination of strict rules and generous prizes to encourage the men at her Dalkeith colliery to keep their homes clean and in good order.[73] In 1818 the Earl of Stafford forbade any North Staffordshire miner living in one of his cottages to keep a dog as a pet.[74] When at the middle of the century free gas was fed to 140 colliery houses at Coltness in Lanarkshire it was intended 'to induce the people to stay at home and read, or pass the evenings with their families, instead of going to public-houses'. To further ensure regular habits the gas was turned off promptly at ten o'clock each evening.[75] Many colliery managers encouraged, demanded even, that their tenants' land should be properly looked after. At Gartsherrie the miners were charged the cost of tending their gardens if they refused to do it themselves.[76] But it was at their boarding-house for single men that the Bairds' paternalistic interference may be seen most clearly. The occupants were allowed to smoke in the evenings but, the rules clearly stated: 'Smoking not permitted *inside* the house during the day. Intemperance and profane or improper language strictly prohibited.' Meals had to be eaten properly: 'No party to be admitted to the eating hall unless his hands, at least, are clean; and no loitering allowed within the rooms after meals have been partaken of.'[77] 'There was no reason', insisted the builder of one model community, 'why they should not have a village where three things could exist successfully ... the absence of drunkenness, the absence of gambling, and the absence of bad language.'[78]

The lives of most colliery tenants were at once less comfortable and less prescribed than those living in houses belonging to aristocratic landowners. The majority of colliery houses were small, poorly designed and in poor repair. In Scotland most were single storey with only a couple of rooms, neither of which were more than about ten to fourteen foot square.[79] The Bairds built some of the best colliery houses in Scotland; yet when they put up new cottages at Kilsyth under the Camprie Hills in the 1850s the bedroom was just fifteen foot by twelve and the total dimensions of the lobby, kitchen and closet combined were only fifteen foot by thirteen.[80] Conditions had improved little by the end of the period. The Medical Officer of Health for Ayrshire reported in 1911 that

> With the exception of two-storey dwellings at Dreghorn, Barrmill and Dalmellington, the houses which have been erected for miners by the coal companies are all one storey, while the accommodation provided consists as a rule of two apartments, namely, a room and a kitchen. There are, however, several rows, such as at Annbank and other collieries, with only one-apartment houses, while in a few isolated instances there are miners' dwellings of more than two apartments each . . . A large number of them are . . . built on the somewhat damp, impervious, cold clay subsoil which generally overlies the coal measures throughout the country; and as the majority of these buildings — practically the whole of the older ones — have no damp-proof courses in the walls, the latter being generally solid without strapping and lathing, they tend to be more or less damp . . .[81]

Colliery houses in Northumberland and Durham were little more spacious during the first three-quarters of the century. 'These cottages are erected', complained a correspondent to the *Lancet,* 'without the slightest regard to the health, comfort, or convenience of the intended occupants. The object appears to be to get the maximum amount of room at the minimum amount of expense. No regard is had as to site, drainage or ventilation.'[82] When the Durham miners' leader John Wilson married in 1862 he and his bride 'started housekeeping in the Long Row at Haswell. The houses were of the back-to-back kind. There were two rooms — the kitchen and the room upstairs, with a straight up ladder as the mode of reaching it.'[83] As time went by, however, colliery houses in the North-East, and in the rest of England, did become a little larger. From just two rooms, the general pattern came to be that of one main room downstairs with a couple of bedrooms upstairs. Where employers provided accommodation in South Wales, it usually consisted of three to four small rooms and three-room houses were also built by colliery owners in South Staffordshire.[84] In North Staffordshire miners working on the Sutherland estate at Ketley lived in cottages with either one, two or three bedrooms.[85] Improvements to Northumberland and Durham homes dated from the mid-1860s and by the end of the period most colliery houses had been built or converted to provide three rooms.[86] A Northumberland miner from New Seaton recalls the home in which he grew up at the turn of the century: a kitchen-cum-living room downstairs and two bedrooms upstairs.[87] A Durham miner from Haswell remembers a similar company house 'containing one large living-room, with a pantry extension. Over the living-room there was one large bedroom. Sizes of families were regarded immaterially, so father, at his own expense, partitioned the upstairs to make two bedrooms.'[88]

Most tenants though were naturally reluctant to invest either time or money in improving houses that did not belong to them. Nor, more surprisingly, was it always convenient for the owners to undertake the necessary repairs. It was said in Durham for example that 'When trade is bad the owners excuse delay on account of expense, and when it is good they say that they have not time' [i.e. that the would-be maintenance men are working in the pit].[89] And although large companies did tend to provide a better than average stock of housing, its maintenance was generally left to agents who saw their main job as being to save their employers both trouble and money. A Durham tenant of the Derwent Iron Company complained in 1864 that the housing manager

> cannot give anybody a civil answer, his insults are base to any one who does not suit him or crosses him. There are some can get anything done to their houses, of alterations or repairs for their comfort, such as call him Mr., and fawn for favour, while others who don't do this or lick the dust at his feet, ask the most necessary repairs done to their houses, get insulted for their audacity, and told in the coarsest language to repair their own houses, and do their own glazing, and if they remonstrate with him he will threaten to shift them away from the place.[90]

The result was that particularly when pits were new or nearing exhaustion or where the trade was specially liable to fluctuation, colliery property was often allowed to fall into very bad disrepair.[91] When the tiles came off the roof of Neddy Rymer's home at Thornley, Durham in 1860, he said that 'Daylight, rain, wind, sleet or snow came in through the crevices.'[92] A few years later the treasurer of the Durham Miners' Association claimed that he 'knew of a boy lying in bed with a broken leg, and an umbrella had to be put up to keep the rain off him . . .'.[93]

If most colliery houses were small, badly designed and not very well maintained, they were often lacking too in even the most basic amenities necessary for good health. Many tenants never had access to a reliable supply of good quality water. Such traditional sources as the well, the rainwater butt and the watercart were often terrifyingly polluted. In the 1870s the water supply of Wylam, a small colliery village seven or eight miles west of Newcastle, still came from two sources: the River Tyne and a local pond. The river received a good deal of sewage, excrement and rubbish from the village, while the pond was 'surrounded with trees and bushes and choked with mud and decaying vegetable matter: the water before reaching the pond, comes from the vicinity of a farmhouse and across a cattle-pasture, in which the chance of pollution is considerable.'[94] Sometimes water was tapped from the colliery workings and stored in a boiler on the surface. An East Scotland Medical Officer of Health remembers that

> In connection with one puzzling outbreak of enteric [typhoid] fever in a mining village, apparently associated with such a supply, I found that the man whose duties included periodical cleansing of the boiler had to go inside, and that he did so whilst wearing boots which might have been polluted by human excrement.[95]

Even when not polluted, the age-old sources of water often broke down in mining districts under the pressure of growing demand. After a visit to the Black Country in 1850 the mines commissioner remarked that 'The ordinary supply from springs is interfered with by the mines, and the population is too large to be properly provided for except through the medium of regular waterworks.'[96] In other districts too the expansion of the industry caused serious dislocation of the water supply. In early and mid-nineteenth century Dunbartonshire,

The mining operations had dried up some deep wells, and part of the surface water was from peaty land. Both quantity and quality were defective, and in dry weather every hole in the moorland which had retained a little water was searched for and emptied, whilst any running pipe yielding a driblet from old mine workings or otherwise could be seen surrounded by a dozen women and children waiting their turn to collect a gallon or two for household use.[97]

The same thing happened in the expanding South Yorkshire coalfield. At Wombwell in 1880 'There was not water except what was got from pumps, wells and gutters, and even the dirty canal was requisitioned to furnish a fluid to wash the coal-begrimed miners.'[98] Pumps were always breaking down. The pump at Low Valley was out of order for three weeks in 1879 and as late as 1898 the pump at Royston Town Well was so weak that it took five minutes to fill one bucket. The result was a queue of a dozen or two women and children at the well every day from five or six in the morning right up until lunchtime.[99]

Slowly the supply of water to colliery houses did begin to improve. Some owners made available the hot waste water from their increasingly powerful winding and pumping engines. But this was obtainable only when the pit was working and often of course the collection point was a good way from the miners' homes. Outside Wolverhampton at Essington the women had to carry their pails a mile or more from the steam engine to their front doors.[100] Slowly too colliery houses were connected to a mains supply. Yet this was no panacea. The waterworks company at Bilston was so inefficient around the middle of the century that 'In some instances the water was on for only about ten minutes during the day, and in others whole streets had none for a week ... The people were taking water from pits and ponds, and wherever they could.'[101] By the 1870s local boards were formed and water companies were functioning in the south of Northumberland and were servicing almost half the colliery villages in Durham.[102] The quality of the water though was not always very high — the Local Board's reservoir at Bedlington was described by one inhabitant as 'a large muck cart stuck upon four brick legs'[103] — and some colliery houses did not have running water installed until the very end of our period.[104] Mains supplies in South Yorkshire were little better. On some October days in 1897 colliery houses at Royston, Ryhill and Havercroft, which drew on Cudworth's four-inch mains, had only one hour's supply of water, a supply which was cut off completely over the Christmas period.[105] In some respects the occupants of colliery houses did even worse than their neighbours living alongside them in other forms of accommodation. The first public water supply was not introduced to the small pit village of Seven Sisters, between the Neath and the Swansea valleys, until the eve of World War I and then the water was connected only to those houses which did not belong to the colliery.[106] But whenever the great day did come, it was well worth waiting for.

> Only those people who have lived in a home without water can imagine what it is like when water was brought to our house. The very thought of it is still wonderful. We were able to turn a tap and have a drink of water when we wanted; able to fill a kettle or saucepan; able to have a wash, without having to go outside to the water tap, take our turn in the long queue and carry the water back to the house. To me and to all the members of our family, running water was a real luxury.[107]

The sanitation of colliery houses, like their water supply, was the responsibility of the employer. It was discharged just as poorly with improvements usually being made only when existing facilities broke down completely. During the first three-quarters of the nineteenth century the tenants of many company houses did not have access even to a communal lavatory. The Midlothian village of Newtongrange was considered 'superior', yet in 1875 it had no closets whatsoever.[108] At the same date houses were being built outside Barnsley at Cudworth without even the room for toilets.[109] In 1872 a reporter from the *Newcastle Weekly Chronicle* found at Thornley village

> a prominent building of much higher architectural pretensions than any of the cottages. It is built of good red brick. After careful inspection and searching inquiry we discover that it was a privy. As such an institution was un-

known in Thornley pit rows before this one was erected, and as the enterprise which promoted its erection has failed to secure imitation, it remains a sort of "9th wonder of the world" to the gaping juveniles who are utterly at a loss to make out its utility.[110]

Not far away at Mickley Square the only privies were those built by the miners themselves.[111] Above and below ground many men had to do the best they could. 'Under such circumstances, fields and woods and hedgerows were resorted to.'[112] It was, as one Durham miner observed in 1863, a quite impossible situation. 'Enclosed as a colliery village is within the limits of one field, and surrounded on every side by other men's property, you have in order to attend to the decencies of life, to render yourself liable to a prosecution for trespass.'[113]

The best that any mining family in a colliery house (or elsewhere for that matter) could reasonably hope for at any time before 1914 was to have an earth closet of its own in the backyard. At Pegswood in Northumberland each house had an earth closet (or 'netty') consisting of a seat perched over a bucket or an ash-pit. Despite the use of disinfectant, these closets stank in hot weather and always attracted swarms of flies.[114] More usually the closet had to be shared. At Wingate, County Durham 'in the back street . . . were rows of earth closets . . ., one closet to every two houses, the neighbours taking somewhat reluctant turns to do the weekly chore'.[115] A great deal depended on the family or families with whom one had to share. One miner's daughter remembers that at Hordern the toilets 'were beautiful clean . . . our part, you know. White wood with a hole . . . used to scrub that, didn't you.'[116] Others were less fortunate. Some toilets were allowed to stay dirty and fall into disrepair. The Medical Officer of Health for Auckland claimed in 1873 to have seen 'privies with no seats, privies with no backs, with no doors, with a seat and no front to it, with half a seat, with no roof, and in some places, the privy instead of a seat had a sort of wooden perch . . .'.[117] Five years later the Medical Officer of Health for Tynemouth complained of middens which 'had accumulated for nearly six months . . .

and had assumed such a state of putrefaction as to be almost unbearable'.[118]

Whatever health risks tenants sharing a toilet may have avoided by living some distance away from the 'netty' were more than made up for by the need to use a chamber pot or to venture out in the middle of winter.

> There was no bathroom, nor toilet except across the way where there was a place called the netty, with open space above and below the door. In all weathers, night and day, it was an ordeal to fulfill the bodily functions.[119]

A miner's daughter brought up outside Seaham remembers that

> we would take a candle in a jar and two of you would go together across the street and into the back yard. Boys playing about would see you and know where you were going. They would get some water from the tap in the street and quietly open the wooden hatch [used for emptying the toilet] and throw the cold water up on to your bottom. What a fright![120]

The disposal of household refuse remained just as rudimentary. Many older colliery houses were quite without any system of drainage. The Medical Officer of Health for Hexham found in 1874 for example that cottages at Wylam, some of which had been built in the eighteenth century, had 'literally no drainage whatever' with the result that 'large stagnant pools of foetid water are found all over the village.'[121] Most colliery houses did have some form of open drainage, generally just by the front door. It was a natural temptation to throw dirty water and other slops straight out into the street. The result may readily be imagined. One Durham miner recalled that on a Saturday morning about 1860 'Mr James Wilson, the manager of the Thornley Colliery, employed a gang of us to clean Quarry Row, and we dug up from two to four feet of cinders, coals and filth through the whole length of the dirty street.'[122]

However the chief means of refuse disposal was always the euphemistically-named ash-pit, a communal receptacle for the receipt of coal-ash, cinders, slops, the contents of chamber pots and anything else that had to be got rid of. At Slalwell in North-West Durham in 1907, 'the only receptacle for refuse for groups of houses is an ancient open brick ash-pit . . . and sometimes it is actually used as a closet by some of the occupiers, notwithstanding its public position.'[123] The ash-pits, like the toilets, were supposed to be scavenged at regular intervals. In some villages they were. From as early as the middle of the century the Cwm Avon Company of Port Talbot was making deductions from the men's pay to meet the cost of sewage and refuse removal.[124] In many villages the pits were certainly not scavenged efficiently. When one colliery opened in Durham in the boom of the early 1870s, profits were poor and the owners built just two rows of cottages which 'are as devoid of sewers as the most neglected villages in Bengal, while in their case neither jackals nor parish dogs exist to perform scavengers' duty'.[125] According to one not unsympathetic observer of the early twentieth-century mining scene,

> The emptying of such privy ashpits is a particularly loath-some process and spectacle . . . The scavenger's duty usually includes attention to privy floors and seats. Where these dry closets are under lock and key, and each convenience duly allocated to certain households, little attention is required. Where, on the other hand, there are no locks and keys, and no allocation, the scavenger's toil is like that of Sisyphus. No sooner has he swept out a privy floor than its defilement by children is resumed, and such places are inevitably in a condition which makes them utterly unusable by an self-respecting adult.[126]

It is possible however to make certain generalisations about the efficiency with which this distasteful job was carried out. Scavenging was done better in the villages than in the towns; it was done better for colliery houses than for privately-owned property; and it was done better as the century progressed. But because it was a service carried out by the owners, scavenging was not carried out when the pit was closed by, for instance, a strike or a lock-out. This was a powerful, if indelicate and rarely discussed, weapon in the employers' industrial armoury.[127]

This leads on to what was perhaps the greatest disadvantage of living in a tied house, the enormous additional power which it afforded to the employer. There was always a good deal of petty favouritism and discrimination. Empty houses in Northumberland were sometimes given to deputies rather than being 'given to workmen in turn as they come to work at the colliery'.[128] At Seaham in 1912 the union claimed that local men were being refused colliery houses, some being told they had too many children, others that they had too few; some were advised to adopt children while another was told that he had no right to adopt a child in order to try to get a house.[129] At many pits it was understood that any tenant of a colliery house was expected, as a matter of course, to send his sons to work underground.[130] Not all discrimination was trivial. Eviction was a powerful threat: in 1866, for example, Derbyshire miners known to have joined the Derbyshire and Nottinghamshire Miners' Association were told to leave both their work and their homes: in one week alone nearly 200 men were given notice by the Staveley and Clay Cross Companies.[131] Eviction was of course most common in Scotland and the North-East of England where a large proportion of miners lived in tied houses. Eviction soured industrial relations, often led to great hardship and sometimes resulted in violence. One example must suffice. After a six-week strike in 1878 warrants were issued on miners in the isolated Slammanan district of Stirlingshire. 'When the officer was engaged serving warrants upon a community of about 120 miners lately in the employment of Mr. John Watson, a large crowd collected, and he was obliged to flee from the place.' The houses of five colliery owners and managers were attacked, 'the windows broken, shrubs and trees destroyed, and furniture and other property smashed to pieces'.[132]

So, like most members of the working class, the tenant of a colliery house had to make do without even the most basic requirements for a healthy, let alone a comfortable, life. But in addition the colliery house tenant was subject to the control of an unsympathetic — and possibly actively hostile — employer. It was this combination of industrial and domestic power which made his position so specially precarious. Yet living in a colliery house did have one great advantage. It was cheaper and more convenient than renting on the open market.[133] Indeed it was far better to have the tenancy of a house than it was to have a rent allowance in lieu. In the North-East for example the allowance was fixed in the 1870s and soon became insufficient to meet the rent of a comparable property on the open market. Nor must well-publicised cases of eviction from colliery houses be allowed to disguise another disadvantage of receiving the rent allowance. Unlike the free house, the rent allowance was paid only while wages were being earned. It was stopped whenever the tenant was sick, injured, on strike or on holiday. The tenant on the other hand usually continued to live in his house in sickness and in health, in peace and in dispute, come what may.

In Scotland and the North-East of England colliery housing created a vicious circle from which it was exceptionally difficult for the miner to escape. The owners tried to build as few houses as possible as cheaply as possible. They were reluctant to leave any house empty because then their investment would be wasted and they would only have to pay a rent allowance instead. The men were equally reluctant to abandon any colliery house no matter how bad it became. Refusal to accept a colliery house could mean the stoppage of the rent allowance, an allowance which in any case was too small to rent a house privately. So it was almost impossible to close down insanitary houses and because the employers owned such a high proportion of the housing stock in Scotland and the North-East, there was little incentive for private enterprise to enter the market to remedy the situation.[134]

A great deal is known then about the minority of miners — a third at the very most — who lived in houses belonging to their employers. On the other hand, next to nothing is known about the housing arrangements of that large number of men, probably two-thirds of the total, who did not live in tied accommodation. While there was a small amount of municipal housing and, in South Wales, a number of philan-thropic building societies, neither made much impact on coal-field housing markets before World War I.[135]

For those without a colliery house there were only three real alternatives: to lodge with another worker; to rent from a private landlord; or to become an owner-occupier. Lodging was common wherever the industry expanded for, almost inevitably, the provision of accommodation failed to keep pace with the influx of new labour. Some of the resultant overcrowding almost defies belief. One of the Children's Employment Commissioners was assured in 1842 that at Coatbridge in Lanarkshire some families in two-room houses were taking as many as fourteen single men as lodgers.[136] With the growth of mining in Fife towards the end of the century some women were packing five lodgers into their tiny attics.[137] It was a national problem: by 1911 almost a tenth of all houses in Durham, Northumberland, Glamorgan and Monmouthshire were being sub-let.[138]

The private tenant, like the surface worker, remains a shadowy figure. (This is not at all surprising for they probably were often one and the same.) But it is clear that the private tenant had to pay more, sometimes far more, than his workmate living in a colliery house. Profiteering was rife wherever the provision of accommodation failed to keep pace with the demand. Housing was very scarce in Lanarkshire in the early 1870s so one enterprising Wishaw landlord put his dwelling up for auction. 'In this way he managed to secure from the scrambling and excited offerers as much as £6 for single apartments.'[139] At the beginning of this century landlords in the rapidly growing South Wales coalfield were able to demand key money and some also compelled their tenants to buy pieces of useless furniture and other items from the shops which they ran as well.[140] It is significant that one of the main targets for rioters at Ebbw Vale, Baryoed and Tredegar in 1911 was said to be house owners who were alleged to be charging inflated rents for their cottages.[141] Private accommodation was especially bad in large towns and in areas of urban sprawl. The mines commissioner concluded of Black Country Wednesbury in 1850 that:

> It is plain that where property is much divided, as in this district, and where there is a constant demand for small houses, every sort of neglect, irregularity, and want of system will prevail in the arrangements for common decency, health and comfort.[142]

In South Wales too the worst sanitation was to be found in the larger towns. The 'Back o' Plough' district of Dowlais was reputed in 1875 to be the most hideous spot in the whole of England and Wales. Its low, poorly ventilated, rubble-built houses were home for many pit labourers. Most houses had four rooms, each about eleven feet square, with floors of small stones, 'up from between which the black ooze squelches when the foot falls unevenly on them — '.[143] Many Yorkshire miners lived in Castleford and right up to the end of the period 70-80 per cent of houses there were still dependent on middens for toilets.[144]

There was one other alternative to living in a colliery house — owner-occupation. It is well known that this became fairly common in the new, relatively high-wage South Wales coalfield where from an early date some miners were building their own homes or buying them in the open market.[145] One observer noted in 1850 for example that at Brynmawr and Nantyglo in Monmouthshire 'A large proportion of the cottages and rows have been ... built ... by workmen out of their savings . . .'[146] Most however came to own their own homes by means of a building club.

A number of miners club together, and with the assistance of a secretary, who is usually an accountant, arrange for a large number of houses to be built in one contract. Each member pays from £10 to £20 down and thereafter monthly instalments of from 10s. to 25s. for each "share", that is, house. When about one-fourth of the cost of each house has been paid in, the club "divides", and each member takes over his house, which is allotted him by ballot, subject to a mortgage which he can pay off gradually like and advance from a building society. The houses are built all the same, so that the allocation by ballot may be fair. Some wealthy gentlemen of the district are appointed trustees, and the building would usually not be financed without their guarantee at the bank.

To meet the financial claims of these clubs men have had to save large sums from their wages to pay for the cost of their houses over a series of from 15 to 25 years, the usual rate of contribution being at the rate of from 15s. to 24s. per lunar month.[147]

During the latter years of the nineteenth century a steady 25 per cent of all new houses in the coalfield were being put up by these clubs of colliery workmen.[148] The result was that 'In parts of South Wales where the club system flourishes hundreds of cottages have been built through the medium of building clubs, and the proportion of owner-occupiers is often very large indeed.'[149] A return made to the Monmouthshire and South Wales Coalowners' Association at the very end of the period confirms this. It shows that at a time when the level of owner-occupation in the country as a whole stood at about seven per cent, more than 19 per cent of all houses in the South Wales coalfield were owned by colliery workmen.[150]

So much is fairly common knowledge. What is less fully appreciated is the extent to which owner-occupation came to be practised in other parts of the country. The odd miner owned his own home in even the smallest and least prosperous coalfields. A collier working at the Alloa colliery in Clackmannanshire in the early 1840s for example owned two houses, one in Alloa and the other in the parish of Clackmannan.[151] It was said in North Staffordshire too that 'great numbers ... in every part of the district have built houses for themselves.'[152] At a large coal and iron works in Kidsgrove 'a considerable number ... have built from their savings excellent and roomy houses, costing 75 l. [£75] each, with enclosed courts and every convenience. Several rows of these houses now form a conspicuous feature in the valley.'[153] When a miner was killed near Ruabon in North Wales in 1863 it was discovered that he lived in a terrace of three houses which he had built for himself and his three children.[154] In South Staffordshire too home ownership was by no means unknown. When the Holly Bank colliery at Essington started to expand at the beginning of this century, more and more miners started to own their own homes, some doing the work themselves and some buying from local builders.[155]

But it was the larger coalfields which provided best the conditions seemingly necessary for fairly widespread working-class home ownership: relatively high earnings, a reasonably stable population, high rents, relatively low construction costs and an absence of middle-class investors and philanthropists.[156] Miners began to buy their own homes in the expanding Yorkshire and Midland coalfields. By the 1860s many South Yorkshire miners had built houses out of their savings and a good number were said to own two or three.[157] In 1880 a land society was started at Unstone, between Sheffield and Chesterfield, in order to enable local miners to purchase housing land cheaply.[158] Indeed at the very end of the period an official of the Derbyshire Miners' Association claimed that 5,000 men — about seven per cent of the country's mining work force — were living in their own homes.[159] Despite the counter-attraction of free colliery housing, it remained the 'burning ambition' of many in Northumberland and Durham and Scotland to own property of their own.[160]

In 1887 there were sixteen miners owning property in the townships of Choppington and Sleekburn.[161] In Durham home ownership seems to have been most common in small places like Butter Knowle, where mining was combined with agriculture, and in the larger towns like West Auckland, where miners' houses along the Darlington Road were said in 1874 to be 'in a word, model working men's dwellings, which may be accounted for by the fact that they are the property of working men'.[162] Cautiously towards the end of the period some North-East co-operative societies began to enter the local housing market. In 1890 the West Stanley society decided to buy property and by 1906 the society's 3,394 members had been able to ballot for 193 houses.[163] Then from the turn of the century the Crook and Neighbourhood Co-operative Corn Mill, Flour and Provision Society had fifteen new £250 houses built and sold them among its members.[164]

Owner-occupation was also more common in Scotland than is generally supposed. In the 1830s many Ayrshire colliers owned smallholdings and round Airdrie 'the men lived . . . in their own cottages'.[165] During the boom of the 1870s union leaders expressly urged their members to become house-owners. The agent of the Larkhall Miners' Mutual Provident Association advised men in company houses to form co-operative building societies 'so that they might be independent of their employers for homes'.[166] The call did not go unheeded for many miners at Crofthead on the Lanarkshire-West Lothian border did own their own houses and it was said that round Carluke 'many have become proprietors of small cottages'.[167] The practice persisted in Scotland to the very end of the period. In Stirlingshire and Dunbartonshire some miners 'in course of years come to build cottages of their own in districts where they regard themselves as permanently employed, or where they propose to live on their savings after their working time has passed.'[168] In Fife too owner-occupation remained the ideal of a substantial minority of the men. At the village of Kelty, for instance, the colliery owned about two-thirds of the houses. But nearly all the rest had been put up by a local building society and were being bought by those who felt able to put down a deposit of 15 to 20 per cent: 'On the outskirts of the village one comes upon small rows of neat, self-contained houses, with flower-beds and grass plots before the doors, and having every appearance of cosiness.'[169]

It is clear that the usual view of nineteenth-century mining settlements as remote, stereotyped, colliery-owned barracks is in need of major revision. Across the country miners made their homes in a whole range of different environments, of which the 'typical' mining village of popular imagination was only one. Yet in the small, culturally (if not geographically) isolated towns and villages of the coalfields, communities were thrown back onto their own resources. The men, together with their wives, did what they were unable to do at the pit. They worked hard — and effectively — to improve the quality of their lives. They bought and improved their homes and created every institution they needed with the exception of the school, the post office and a number of specialist shops.

> There can be no question at all that these [Durham] villages (most of which assumed their present size during the middle years of the nineteenth century) owe all their horror to the colliery-owners who built them, and are their redeeming features to the organisational energies of the people who live in them.[170]

It was a remarkable achievement and one which goes far towards invalidating the stereotype of the irresponsible and thriftless miner.

5

The Miner at Home

The miner's domestic arrangements have received an almost uniformly bad press. It is said that he was utterly irresponsible. He married young and went on to produce a child a year until — and even after — his wife was exhausted and his home impossibly overcrowded. He thought nothing of the future and lived only for the present, spending as much as he could on drink and gambling. The clergy, the inspectorate and the employers all fulminated against the miner's fecklessness. 'To economy,' regretted the Reverend Gisborne in 1798, 'he is, in general, an utter stranger.'[1] The first mines' inspector, Hugh Seymour Tremenheere, regularly attacked the extravagance of miners and other workers.

> Poultry, especially geese and ducks; the earliest and choicest vegetables (*e.g.* asparagus, green peas, and new potatoes, when they first appear in the market); occasionally even port wine, drunk out of tumblers and basins; beer and spirits in great quantities; meat in abundance, extravagantly cooked; excursions in carts and cars, gambling, &c., are the well-known objects upon which their money is squandered.[2]

A boom in the coal industry always led to criticism. In 1872, for instance, the *Colliery Guardian* attacked the improvidence of the miners in Fife and Clackmannan: 'Provident habits being very rare among the mining population, they are spending their present abundance in useless indulgences and intemperance.'[3] Even miners and those sympathetic to them have tended to agree. Neddy Rymer, an agitator with first hand knowledge of every major British coalfield, admitted that though 'the miners worked hard, . . . [they] spent their

112

money quickly.'[4] The historian of the Derbyshire miners believed that 'There was, no doubt, a streak of Epicureanism in
the miner's philosophy, which has not altogether disappeared.
Never knowing when bad times, or even fatal accident, would
befall him, he tended to enjoy himself whilst the could.'[5] Dr
Challinor agrees:

> Miners' social life was inevitably affected by living amid
> squalor, dirt and degradation. While mental improvement
> societies and other educational bodies endeavoured to en-
> lighten them, such activities appear to have been very much
> a minority interest. Most men seem to have neither the time
> nor the inclination. Their main purpose appears to have
> been drinking, often heavy drinking . . . A union leader
> criticised men "whose minds never soared any higher than
> the brim of beerpot".[6]

There is no doubt that the nature of the miner's work, the
way in which he was paid and the type of house in which he
lived all caused very great disruption to domestic life. Indeed,
perhaps in no other industry did the way in which a man
earned his living have a more profound effect on the way in
which he and his family lived at home. The pit intruded everywhere. Even if he should be lucky enough to avoid serious
injury and crippling disease, the miner and those around him
had to learn to live with his tiredness, his aches and pains, his
ruptures and his rheumatism. Putters always had sores and
scabs on their backs from banging them on the rooves underground. 'It was an ordeal at weekends sitting throughout a
Sunday Church Service all the while pressing one's back
against the pew to obtain relief.'[7] Even the journey to and
from work was not without its hazards. 'Men's feet were often
covered in blisters by the time they had walked three to four
miles from the pit in "clarty" [muddy] socks or heavy
boots.'[8] Those travelling to and from work by train were not
necessarily any better off. Men working at the Dulais, Onllwyn
and Seven Sisters collieries in the Dulais Valley protested
against the 'trying inconveniences' of having to wait around
for late trains, often while in clothes that were still wet from
the pit.[9] There were other problems. At some collieries the
men were expected to make up their explosives at home. The

results could be catastrophic. In the summer of 1879, a Derbyshire collier took a lighted pipe into the room where he kept his gunpowder. He 'was frightfully burned, and his wife less seriously so, and the cottage was completely wrecked.'[10]

The pit intruded in other ways too. Miners started work earlier than most. In nearly every coalfield the first men down, the officials, the hewers and the trappers, began between 4 a.m. and 6 a.m., with the oncost workers following an hour or two later.[11] In Northumberland and Durham, where one long haulage shift serviced two hewing shifts, the first shift of hewers went down even earlier. When John Wilson began hewing at Sherburn Hill in 1854, 'The fore-shift man went down at the latest at 2 a.m., but some went much before that hour. If a man in back [shift] was later than 6 or 6.30 a.m. he was late.'[12] Starting at these times meant of course getting up very early indeed. When sixteen year old Joseph Halliday began at South Hetton colliery just after World War I, he 'had to rise at 2.45 a.m., leave at 3.15 and walk for half-an-hour' to be at the pit by a quarter to four.[13] Few miners would have had a longer journey to work than this in the first half of the nineteenth century but with the expansion of the Midland, South Yorkshire and South Wales coalfields more and more pits were opened away from established centres of population.[14] For years hundreds of Barnsley miners used to catch the 5.15 a.m. train out to the Carlton and Monk Bretton collieries.[15] By the end of the century the South Wales coalfield was criss-crossed with railways, the Taff Vale line alone running twenty-three workmen's trains a day in each direction, and carrying its average passenger at least sixteen miles.[16] With the lengthening of his journey to work so the miner in the newer coalfields lost much of the benefit which he might otherwise have gained from the progressive shortening of his working hours.

It was never easy to get up in the middle of the night to go to work. Many miners owned clocks and watches but in the villages most preferred to rely on the 'caller' or 'knocker-up', a retired man who was paid to rouse the workmen according to the time scratched on a slate by the side of the front door.[17] The arrangement did not always work smoothly.

In some villages, complained the hewers, the calls were made to suit the deputies with the result that everybody was woken up at 1.30 a.m.[18] Naturally it was difficult for the caller to wake the husband while leaving the wife and children undisturbed. Some callers were quite unsuited to their job. In one Durham village in the early years of this century, for example,

> He would knock at the door with a knobby stick, and shout "caller". He would continue to knock until you answered ... I remember one caller was very deaf and would go on knocking until someone opened the door.[19]

Even experienced workers did not necessarily find it easy to get up so early. A Hetton miner was ...

> . . . aroused from a deep sleep at the ungodly hour of 2.30 a.m. to go on first shift. He was on the point of leaving home, grabbed what he thought was his "bait tin" wherein his food was always packed, tucked it under his arm and set off. Later reaching the coal face one thousand feet down and two miles in, he went to put his tin in a safe place, then found to his astonishment it was not what he thought it was, but his alarm clock![20]

Yet the inconvenience of getting up early to start work was nothing compared to the domestic disruption caused by the introduction of multiple-shift working. Mid-century improvements in haulage and winding made it possible to move more coal to the surface and increasing capital investment made it more desirable to try to spread fixed costs over a greater output. Double-shift working was always the usual method of working in North Staffordshire and it became increasingly common elsewhere in the more modern, capital-intensive pits.[21] Between 1888 and 1905 many large, heavily capitalised South Yorkshire collieries switched to double-shift working and in the Midlands the new system was adopted at many of the new 'top-hard' collieries until in 1913 about a seventh of all Derbyshire collieries were on double-shift.[22] At some North-Eastern pits the introduction of the Eight Hour Act at the end of the period led to the old practice of one haulage shift servicing two hewing shifts giving way to a more complex system whereby two haulage shifts serviced three shifts of hewers. So in 1910 at the Hewarth colliery in Northumberland the first shift of hewers worked from 4 a.m. to 11 a.m., the second shift from 9.30 a.m. until 4.30 p.m. and the third shift of hewers from 3 p.m. until 10 p.m. Meanwhile the first winding shift worked from 6 a.m. until 2 p.m. and the second winding shift went down at 1 p.m. and came up at 9 p.m.[23]

Multiple-shift working was always unpopular. It meant the hewers had to share their stalls; it was felt to be more dangerous; it was thought likely to produce a glut of coal;[24] and it made it difficult to sleep during the day.

> There were no roaring loud-speakers in those day, but there were children playing in the street and hawkers with voices that sounded to be shaking the bed I tossed about on.[25]

Above all it was difficult to get into any sort of routine. The first week Joseph Halliday worked at South Hetton his hours were 3.45 a.m. to 11 a.m. Monday to Friday, and from midnight until 6 a.m. on the Saturday. The following week he worked 'that loveliest shift of all, 9.30 a.m. until 4.30 p.m. with Saturday 6.0 a.m. until 12 noon'. But soon he had to work 'the "old men's shift", three o'clock in the afternoon till ten at night'.[26] This was particularly unpopular.

> The elderly man with a steady bit of gardening in the mornings, that had about them crispness and fresh air-ness, was quite content with the afternoon and evening toiling, at the end of which was a good night's rest. The young man, however, wanted . . . his evenings free and so it was not unknown for some who, faced with a week of evening duty, would approach older men not nearly so gay nor giddy and offer a sum of money, thus arranging a swop of shifts.[27]

Whatever shift he was on, the miner's homecoming disrupted the whole of his tiny house. The miner expected a hot meal and a hot bath. With pithead baths almost unheard of, he 'usually performs his ablutions in a tub placed in front of the fire, while the other members of the family pursue their ordinary avocations.' Often 'four or more are awaiting their turns for the tub, there being perhaps a father and two of his sons, and one or two lodgers, all working in the same pit.'[28] How much greater was the dislocation when, as often happened later in the century, miners living in the same house were on different shifts. In a family with working sons and perhaps a lodger, there was always someone to get off to work, someone to feed, someone to prepare a bath for and someone's clothes to wash and dry. Multiple-shift working, protested the Lancashire miners' leader Thomas Halliday, was 'an attack on the happiness and comfort of the men'.[29] He ought not to have forgotten the women for as has often been pointed out, 'The very nature of pit work made most women slaves, wives and daughters all. Shifts split up the family so that men would be coming in at all hours of the day, waiting for the bath-tin and the water and a woman to wash their backs.'[30]

Perhaps even more disruptive than early rising and the introduction of multiple shift workings was the unpredictability of the miners' earnings. As was made clear in an earlier chapter, even the most settled and steady of men could never be sure how much he would bring home the next day, let alone the next month or the next year. This uncertainty made planning the family budget a task bordering on the impossible. Uncertainty, however, was not the only problem. Almost as important in determining the quality of family life was the regularity with which wages were paid. During the first three-quarters of the nineteenth century the vast majority of British miners were paid fortnightly or even less frequently. Warwickshire, Lancashire and North-East miners, for example, were paid every two weeks while most men working in the Forest of Dean, in the Midlands and throughout the whole of Wales were paid only monthly.[31] It could be even longer. The owners of one St Helen's colliery in the 1860s were keeping back wages for five weeks with the result that 'it got spread about that the colliery was broken, and the shopkeepers were afraid of trusting them any longer.'[32] Not surprisingly many families found it impossible to manage for so long and were forced into truck. The employer might allow an advance on the next pay, but only on condition that it was spent at one of his shops.[33] At pits belonging to some South Wales iron companies in the 1870s and 80s, the workmen would be graciously permitted to draw a portion only of their wages, every week, the remainder was paid at varying intervals, sometimes as long as twelve or fourteen weeks.[34]

Slowly, weekly pays did become the rule. While it is true that as late as 1900 the members of the Durham Miners' Association decisively rejected the adoption of weekly pays, by the turn of the century most other miners in England and Wales were being paid every week.[35] This was an important change for it made it easier — though never really easy — for the miner and his wife to plan the family budget.

Crucial too in determining how wages were spent was the time and place at which the miner was paid. A Friday evening pay-out gave the women all day Saturday to shop. In the North-East wages were paid at the pit every other Friday evening and it was customary to spend the next day, the pay (or baff) Saturday, in the nearest market town. Whole families flocked into Newcastle, Durham, Sunderland and other towns to drink, to shop and to see the sights.[36] So, on pay Saturdays around the middle of the century, the owners of the Evenwood colliery in south-west Durham used to send a two-horse wagon into Bishop Auckland 'conveying the pitmen's wives in order to make their marketings.'[37] But in the more backward districts, where few pits had office accommodation and where banking facilities were perhaps rather less adequate, the legislation of 1842 appeared to have little effect and wages continued to be paid in the pub, often late on a Saturday night.[38] This encouraged drinking and made careful shopping almost impossible. At the Russells Hall colliery near Dudley in the early 1860s, three or four butties were in the habit of paying the men their wages in the 'Horse and Jockey' at Lower Gornal, about a mile from the pit and still further from Dudley market where the miners and their wives wanted to shop. The butties would not even begin paying until 9 or 10 p.m. 'which prevents us attending the market until near midnight, or compels us to be huckstered by small vendors in the neighbourhood where we reside.'[39]

Slowly, however, men in backward districts like the Black Country did begin to be paid more regularly at the colliery, in the same way as their fellow-miners in other parts of the country. But nowhere was anything done to overcome the serious problems arising from the variability of miners' earnings. No more in the coal industry than elsewhere did 'Cyclical economic depression [prove] to be [an] inducement to the development of financial prudence and to the exercise of forethought in the family economy.'[40] Nor was anything done to reduce the very great strain which the dirt, the danger and the unsocial hours of pit work placed upon family life. Indeed the difficulties became exacerbated as double, and even triple-shift working spread through the more important coal-fields during the latter years of the nineteenth century.

As if all this were not enough, home life was made more difficult still by the small, dark, damp, insanitary buildings in which so many working-class families had to live. Minute windows and the lack of gas or electric lighting made most colliery houses dull and gloomy. Even at the end of the century 'Life seemed always to be lived in shadows and darkness.'[41] This was not only depressing; it could also be dangerous. At Ned Cowen's home in the Durham village of Bewicke Main there was no handrail on the stairs and only oil lamps and candles for lighting. One day Ned and his sister, Martha, fell headlong down the stairs. Martha fractured her spine and died soon afterwards.[42] Even the advantages which the miner enjoyed compared to other members of the working-class, could reduce his chances of a healthy home life. The free coal allowance enabled those who received it to keep warm and eat cooked meals.[43] But in Scotland at least the coal was generally stored indoors, often under a bed, and the poor ventilation of nearly all colliery houses meant that life was lived in a fog.[44] And, 'What can be more injurious to a man's constitution', asked one miner, 'than to rise from his warm bed in an almost perfect state of fever by being shut up in a small room with no ventilation whatever — for such is the dwelling of thousands of north country miners.'[45]

Far more serious than the dark, the damp and the heat were the sanitary defects which loomed so large in the last chapter. Like the towns, most nineteenth-century mining settlements were extremely unhealthy places in which to live. Polluted water supplies and inefficient waste and refuse disposal resulted in the rapid spread of digestive complaints like diarrhoea and of zymotic diseases like typhoid (enteric fever), scarlet fever, measles and whooping cough.[46]

> Unscavenged ash-pits and privy middens were serious factors in the production and conveyance of disease, particularly by the breeding of flies, which, with their feet dipped in filth, invade the houses and settle on articles of diet, including milk and jam and butter, and on children's faces after food. Also dogs and cats may rake amongst the refuse and afterwards play with the children, or contaminate food or milk, often standing exposed on a kitchen table.[47]

Hot dry weather always aggravated the problems. It made the drainage even more offensive than usual and dried up the wells and streams. Throughout the century 'autumnal diarrhoea' swept through the colliery villages.[48] Periodic illness was a way of life – and death. 'If we had missed our annual fever, diptheria, and other epidemics,' observed the mining MP, Jack Lawson, 'we should have thought that something had gone wrong with the seasons.'[49]

All the miner's domestic difficulties were aggravated by the fact that more often than not his home was seriously overcrowded. It is common to attribute this to his irresponsibility in marrying young and having more children than he could cope with. Certainly it is not difficult to find examples of miners fathering ten, twelve, or even more children. Alexander McDonald was the eldest of seven; the Cumberland trade unionist, Andrew Sharp, the youngest of nine; William Harvey of Derbyshire the youngest of five; and Tom (later Lord) Williams was one of fourteen children.[50] The view that miners 'displayed a fondness for young brides, and also for high fertility' dies hard.[51]

Yet the accusation turns out to be little more than one of the standard, and often unsubstantiated, middle-class complaints about the working-class. From the countless criticisms made of the miners, it would be easy to imagine that they were getting married almost before they had left adolescence. Nothing could be further from the truth. Despite the fact that miners attained their maximum earnings while still young, only a few ever married while in their teens. Through-

out the period North-Eastern miners, who were the best paid
and presumably therefore the most likely to marry young,
married 'at round about twenty to twenty-three years of
age' or even a little later.[52] (The average age of first marriage
for all nineteenth-century English bachelors was not much
higher — between 25.5 and 26.6.[53]) Miners in other coalfields
married when they were slightly older. In 1884-5 the average
age of marriage of all miners (and their wives) stood at 24.06
(and 22.46), compared to 24.38 (23.43) for textile workers,
24.92 (24.31) for shoemakers and tailors, and 25.56 (23.66)
for labourers.[54] So although miners married a year or so
younger than other workers, most waited until they were in
their middle twenties. Search as one will, the age of marriage
provides little evidence of the miners' almost proverbial irresponsibility.

Once married, the miners' carelessness is supposed to have
emerged in the large size of their families: 'If there were six
or eight in the family and all too young to be earning anything, it made an unbelievable difference to the kind of
shelter and food that the family could afford. Even when
there was a small family, it needed steady work, careful spending and some extra money coming in from somewhere before
there could be any real comfort.'[55] This is a difficult area to
discuss for almost nothing is known about sexual behaviour
in the coalfields. But it is important not to exaggerate either
the size of the miners' families or the detrimental effect which
having several children could have on the domestic economy.
In fact there is no reliable evidence as to precisely how large
mining families were during most of the nineteenth century
— and it is impossible of course to rely on any historian who
simply equates family size with the number of people living
in a house. However, the 1911 Census of Marriage and
Fertility does provide a comprehensive picture of early twentieth-century marital behaviour. It shows that while at this
date the typical mining couple had 4.23 children, way above
the national average of 3.53, families in several other important occupational groups had larger families still to bring up:
building labourers had 4.35 children; iron and steel workers,
4.36; general labourers, 4.41; and agricultural labourers, 4.51
— eight per cent more than were to be found in the average
mining family.[56]

It is possible also to calculate the number of dependent, school-age children belonging to each colliery family. This is particularly fortuitous for in some respects the number of dependent children provides a more sensitive indicator of pressure on the family economy than does the total number of children, whether grown up or not. School-age children needed more time and attention and were unlikely to be able to contribute very much towards their keep. Estimates of the average number of children dependent on each miner may be derived from three main sources: newspaper stories following major disasters; the mines' inspectors' reports and, from about 1870 onwards, the extensive records of the miners' permanent relief funds. These suggest that at the beginning of the nineteenth century most couples had to look after between three and three and a half children; that by the middle of the century the number had fallen to about two and a half and that by the end of the century the typical mining family had less than two children under the age of thirteen.[57] Over the course of the century the number of school-age children in each family fell by more than a third, from something over three to well under two. This was a clear trend and most important. It eased the financial pressure, helped to make the overcrowding problem a little less acute and gave the wife a little more time to herself.

It is apparent that the reason for the overcrowding of mining homes is not as obvious as it may seem at first sight. The expansion of the coal industry created a disproportionate demand for young workmen so that the age and sex structures of nearly all nineteenth-century mining communities were sharply out of line with the rest of the country. In mid-century Merthyr, for example, about two-thirds of the population was aged between fifteen and thirty-five (against a national average of 56 per cent) of whom almost 60 per cent were men.[58] The result is easily imagined. In parts of Monmouthshire and East Glamorgan in 1911 the proportion of married women in the child-bearing age range stood at more than twice the national average.[59] So it is not surprising that all over the country nineteenth-century mining settlements also contained a large number of children.[60]

The overcrowding problem was caused less by the miner's

irresponsibility in marrying young and fathering large families than by having to live in one or two rooms, by the demographic imbalance of mining communities, and by the widespread practice of taking in lodgers. Of course, it is easy to criticise the miner for choosing to share his already small home. But many a hard-pressed housewife must have welcomed any extra help with the family budget. At the end of the century lodgers at Kelty in Fife were paying twelve shillings a week for bed, food (which cost only three shillings to buy), washing and mending, as much as the head of the house could hope to make in two shifts.[61] Some families had no choice in the matter. For example in the North-East of England Pease and Partners required their tenants to accept lodgers as a condition of tenancy.[62] Often it was a matter of giving orphans and relatives a roof over their head. 'If a lad is fatherless,' observed a Sheffield vicar in the 1840s, 'the others always look at him, and make out for him.'[63] As in other working-class communities, it was common for 'The newly married couple . . . [to] set up house in one upstairs room of their parents' home' and use the money saved on rent to buy furniture.[64] Indeed some of the very worst cases of overcrowding do seem to have occurred as a result of offering shelter to relatives. At the end of the century, to take just one example, a forty-three year old miner, his eighty-five year old mother, his daughter and his grandchild were all found sleeping in one room at Darlaston in the Black Country.[65]

Whatever the cause, the result was the same. The miner's home was cramped and probably crowded. In every coalfield it was the rule to find five people or more living in each house.[66] The number could climb much higher. In 1846 the average cottage at Maesteg in Glamorgan housed twelve people.[67] The inconvenience of living in these overcrowded conditions is easily imagined. 'What do you think', asked a North-East miner in 1863, 'of a poor man and wife with five or seven children in one room, twelve or fifteen feet square, and that one room must answer the purpose of bakehouse, washhouse, laundry, pantry, and sometimes coalhouse besides.'[68] He might have added, without exaggeration, that this one room would also have to serve as dining room, bedroom and sickroom. Depending on the number of inhabitants, it was usual to keep one or more beds in the living room. In prosperous districts like Northumberland and Durham these might take the form of a folding bed, known as a cheffonier, which could be converted into a wardrobe during the day.[69] The young Billy Brown remembered the sleeping arrangements in his parents' colliery house at North Seaton in Northumberland around the turn of the century. Billy and his three brothers slept in one upstairs bedroom, his two sisters slept in the adjoining room and his parents slept downstairs in the 'kitchen'.[70]

The cramped conditions exacerbated the lack of amenities to make even the most common domestic activity into a battle of tenacity and ingenuity. Wash-days were bad for the health and worse for the temper. Even in the few cases where new colliery houses were built with separate washing accommodation, the wash-house was sometimes most badly sited. In parts of Stirlingshire and Dunbarton for instance it was built alongside the 'foul smelling midden'.[71] The misery of wash-day is still vividly remembered today. According to a miner's daughter,

> Washing days were the greatest dread in those days. I remembered our clothes were made of strong heavy stuffs, like worsteads or woolens, for endurance as they had to be handed down. They were hard to wash and harder still to dry . . . We had a long brass line hung above the fireplace on which towels were always drying. In a big family like ours it was a problem to keep towels dry for use.[72]

Drying the washing was the greatest problem. A Durham miner remembers wash day as 'the Devil's birthday':

> a living room reeking with steam, and clothes drying out, pegged on one or two lines stretched across the room. Often pit clothes drying out on a line stretched across the front of the fire place and all in the dim light of an oil lamp sometimes augmented by a lighted candle.[73]

In bad weather the clothes 'hung in the small kitchens for nearly a week, and tempers were frayed to breaking point, trying to dodge damp clothes'.[74] In good weather clothes were hung up outside 'and your mother would have her washing out when the midden cart came and all the washing had to come in'.[75]

Of a' the plagues a poor man meets,
Alang life's weary way,
There's nyen amang them a' that beats
A rainy weshin' day.
And let that day come when it may,
It a'ways is maw [my] care,
Before aw [I] break maw fast, te pray
It may be fine and fair.
For it's thump! thump! souse! souse!
Scrub! scrub away!
There's nowt but glumpin' [sulking] i' the hoose,
Upon a weshin' day.[76]

If overcrowding and lack of space made it hard to cope with such ordinary, everyday activities as doing the washing, they made it virtually impossible to deal with domestic emergencies. 'When all the members of the family were in good health the sleeping accommodation was often a worry,' recalls a Fife miner, 'but if anyone was ill, it was a problem and shake-downs had to be resorted to.'[77] Isolation was quite out of the question. 'A miner's wife is confined and there is but one room for her accommodation. How in the name of common sense and common decency is this allowed to be?'[78] Or again, 'Think of a miner receiving an injury in the mine by one of those accidents to which he is liable at his daily labour. How can he be kept quiet in this one room?'[79] In 1898 a doctor called to attend a mining family at Clay Cross in Derbyshire. He found that the mother, in a pathetic attempt to prevent the spread of infection in her overcrowded home, had placed three children with scarlatina in one end of a bed and three with typhoid in the other.[80]

There is not the slightest doubt that the nineteenth-century miner and his wife faced enormous difficulties in their life together. The cramped, unhealthy home, the lack of privacy, the ways in which wages were paid, the early starts, the comings and goings, indeed the very fact of working at the pit, made it far from easy to enjoy a stable and comfortable

home life. Yet what is so striking is not the scale of the disruption but rather the way in which ordinary mining families
attempted to overcome these difficulties. One way of building
a more settled family life was to try to stay in one area. It is
always said of course that one way in which the miners' irresponsibility manifested itself was in their constant movement
from pit to pit and from district to district. According to this
view, trade fluctuations, the opening and closing of pits, the
danger of the job, the annual bond, the paucity of amenities
in mining settlements and the lack of loyalty felt towards
colliery-owned housing, all meant that early nineteenth-
century miners 'tended to live their lives from day to day
only, and they adopted a careless attitude to the normal res-
ponsibilities of life including that of providing a settled home
for their families.'[81]

Certainly the first half of the century did see a very great
migration between — and especially within — the coalfields. It
has been estimated that something like a tenth of all colliers
moved house every year, a figure which rose occasionally to
as high as a third in Northumberland and Durham where the
annual bond was in operation and where it was the owners'
responsibility to move the furniture of incoming workers free
of charge.[82] A miner with experience of working in the North-
East remembered that

> Those miners who were dissatisfied with one colliery would
> get bound at another, so that shifting chattels and family
> became necessary. In this way many thousands of families
> would change districts annually, and in April and May the
> whole mining districts were alive from side to centre with
> loaded vehicles from BIG WAGGONS TO "CUDDY"
> CARTS.[83]

But even in the first half of the century such large-scale
family migration was the exception. And it does seem that
as large pits came to have longer working lives, as the annual
bond was abolished, as local rail communications improved,
as mining settlements began to acquire a few more facilities
and as home ownership became more common, so fewer and
fewer families cared to face the upheaval of moving house.
This is not to say that long-distance migration no longer took

place. Depression, the opening of a new coalfield or the extension of an old one, all continued to uproot large numbers of men.[84] Thus at the start of the twentieth century the Doncaster coalfield drew miners from Durham, South Wales and many other places.[85] It was South Wales itself though which was the biggest draw. Thousands upon thousands of men poured into the coalfield from rural Wales, from Gloucestershire, Cornwall, Somerset and the West Midlands. Between 1901 and 1910, nearly 100,000 new workers came into the South Wales coalfield: '*Anybody* could get a job in the mines', recalled an old collier.

Llwynypia Colliery was all Bristol people. Boys would come here and then go home Easter-time, Whitsun-time, and Christmas-time, and then they would bring half-a-dozen or more back with them. Going to the top of the pit, seeing the boss: "How many have you got?" "I've brought six back with me." "All right, start in the morning."[86]

There is no doubt that once they had become miners, many men were becoming increasingly reluctant to keep uprooting themselves and their families. Those who did continue to move were often young bachelors looking for better pay, easier work, more freedom or just a change.[87] When a family man wanted to change pits, he was far more likely to look for one within commuting distance of his house. This was quite feasible, for, as has been made clear, few villages (even in Northumberland and Durham) were ever totally dependent on a single colliery. Almost always there were at least one or two others within reasonable commuting distance. It has been found that even in years of rapid expansion, a surprisingly large number of Lanarkshire miners remained loyal to one village. A fifth of the mineworkers living in Coatbridge and Gartsherrie in 1841, and more than half those in Low Quarter and Larkhall/Millheugh, were still there in 1851.[88] By the time that the South Wales coalfield underwent its great expansion in the second half of the century, the balance between a constantly shifting extractive industry and an essentially stable mining population was being maintained by the railway system.

Thus every effort was made to encourage *daily* work-journeys by rail in any area where mining was tending to gravitate away from existing settlement ... The effects of such work-journeys on the social and economic development of the coalfield were of great importance. They made possible a flexible relationship between the needs of industry and the distribution of population in the coalfield, so enabling the mining settlements to build up an element of stability so valuable for social and community development.[89]

The miner's lot improved then as he and his family became more settled and as he and his wife came to have fewer young children to look after. It improved too because, as has been seen, his disposable income grew significantly during the course of the nineteenth century. Truck declined, fining became less severe and average real earnings rose about three-fold. These were all important determinants of social life in colliery communities. Yet they affected the wives and daughters of the miners far less than might perhaps be imagined. So long as the miner's job remained hard, dirty, exhausting and little affected by technological change, so the miner's wife found it almost impossible to escape from her traditional, supportive, maternal role. As a visitor to the Fife coalfield remarked in 1902:

> In the miner's world the man is the keystone to the household arch. Woman's place is to support and buttress him from every side ... Their slavery to the men was almost universal throughout the district. The men were looked upon as the wage earners, and the lives of the women were given up to making them comfortable.[90]

The miner's work was done as soon as he got home and he expected to be waited on hand and foot. Even the newest recruits to the industry knew the score. Here is the young Joseph Halliday after his first day at South Hetton colliery.

> Putting my pit-bottle and tin on the table I divested myself of jacket and waistcoat, took off my toe-capped, hob-nailed boots, placed them in a heap on the concrete yard outside, washed my hands at the kitchen sink then sat straight down

in a corner seat while mother put before me a lovely 'pot-
pie', [meat-pudding] . . . After the meal and by the time I
had read the daily paper Mother had got the bath and hot
water ready. Off went my blue pit socks, top trousers . . .
My flannel shirt and vest joined them in a pile which
Mother carried outside to shake, knock and air.

His mother washed him, he went to bed and he was called
when tea was ready.[91] In another part of Durham.

It was the job of the girls of the family to ease the lives
of the miners by having hot water ready to fill the tin bath,
and after the bath we had to wash out the flappers and
socks and put them to dry.[92]

Girls and young women helped their mothers with many of
the household chores. They cleaned and greased the pit boots,
filled the boiler with coal, helped with the washing and
scrubbed anything made of wood — 'baking boards, rolling
pins, potatoe mashers and broom handles . . .' They had to
earn their keep. 'Daughters', remembers a girl brought up in
Durham, 'had to work very hard in a family of pit workers.'[93]

Despite being more settled, despite having more money to
spend and fewer children to look after, the life of a miner's
wife, like that of other married working-class women, remained one long round of chores and drudgery.[94] At the end
of the period, as at the beginning, there was the washing to
do, meals to prepare, bread to bake, hot water to boil,
cleaning to do, boots to repair, shopping to get and children
to care for.[95]

The women work very hard — too hard — trying to cheat
the greyness that is outside by a clean and cheerful show
within. They age themselves before they should because of
this continual cleaning and polishing.[96]

There was no escape. The wife might be responsible for
buying the food and clothing but shopping was no release;
it was a chore.[97] At the truck shop it was an ordeal: 'Wives
and children were compelled from an early hour to wait their
turn at the stores . . .' A woman whose husband was employed
by the Monkland Iron and Steel Company in Lanarkshire

protested that 'When they give you your article they pitch
it to you like a dog. They are sure of their money and know
that you must have your line [order].'[98] Nor is there any
evidence to suggest that the decline of truck and the opening
of other shops led to a greater enjoyment of shopping. There
was too much to buy with too little money. Even in relatively
high-wage coalfields like Durham, caution was the key word.

> I remember going shopping with mother and a thing would
> be marked 6¾d. Now the farthing was being phased out,
> so mother reckoned if you could not get your farthing
> change why mark things in farthings. So she decided she
> would have her money's worth and asked for a farthing's
> worth of pins in lieu of her farthing.[99]

Nor did going out to work offer any sort of release. In
working-class families taking a job did not mean the sub-
stitution of one form of menial labour for another. It meant
rather that the housewife would be expected to do her new
job as well as performing all her usual domestic duties. There
was little enough choice of job in the coalfields, even before
women were excluded from underground work in 1842.
There was factory work to be had in the textile industries on
either side of the Pennines and a little pit work in parts of
Lancashire, the West Midlands, Pembrokeshire and the East
of Scotland. Otherwise it was a choice between doing sewing,
taking in washing, helping in the fields, going into service or,
as a very last resort, 'hawking things about the country'.[100]
Not surprisingly most women seem to have been happy
enough to give up work when they got married or as soon
as they started a family. The few who did work were generally
those with a number of small children yet with no grown-up
sons to boost the family income.[101] But by the time a mother
had paid for her children to be looked after, it was often
hardly worth while going out to work. In 1844, for instance,
a mother of four was earning 7s. (35 pence) a week at the
Pencaitland colliery in West Lothian. But she had to put up
with the damage which the children did at home, she had to
pay a shilling (5 pence) a week for someone to do the family
washing and 2s. 6d. (13 pence) to a woman for looking after
'the younger bairns' (and she had to get them to her by 4

a.m.).[102] So few miners' wives combined work with marriage. Thus in 1861 only thirty-seven of the 358 living at Larkhall in Lanarkshire, and a mere three of the 680 wives at Coatbridge, were out at work.[103] Most women found it more than enough to look after their husbands, clean their homes, do their chores and see to the children. They had, if anything, an even worse time of it than their husbands. As girls they were expected to work hard around the house; as women they 'were considered old as soon as they passed child bearing age and were treated as old people'.[104]

Control over such domestic activities as washing and shopping, important though they were, did nothing to alter traditional relationships within the family. It was the husband, with perhaps his sons, who bought resources into the house. The balance of power within the home was likely to be upset only when the wife too was able to make a permanent and valued contribution to the family income.[105] With the domestic role of women changing so slightly, it is not surprising to discover that there was little change in the treatment of children. Again, however, the evidence is scarce and it is difficult to write with any degree of certainty. But it does seem that in some respects tiny babies always tended to be treated like miniature adults. Whether breast-fed or brought up on cow's milk, they were weaned very young and put onto the solid foods eaten by the rest of the family. In parts of South Yorkshire babies were said to be fed on pickled onions and tinned salmon. This was convenient no doubt and enabled parents to boast of their children's precocious capacity, but it can have done the babies' digestion no good at all.[106]

Parental indifference and ignorance persisted in working-class communities well into the nineteenth century.[107] There is no evidence that attitudes in the coalfields were any better or any worse than those elsewhere. Infant illnesses tended to be ignored (babies with bronchitis were sometimes taken out late at night in the middle of winter) or treated with gin and quack medicine. One expert claimed in 1842 that the worst problem in the South Staffordshire coalfield was that mothers treated their children with things like 'Godfrey's Cordial', a dangerous mixture of treacle and opium. 'Medical men seldom see the children until they are benumbed and stupified with opiates.'[108] Vaccination was distrusted; a Leeds miner complained:

> Is it not disgusting to be compelled to take your innocent child to a doctor, before it is three months old, to have its arm scratched and poisoned with the dirty vaccine matter, and thereby rend it liable to become subject to some terrible disease, and in many cases at the risk of its life?[109]

Accordingly disease was rampant. Diarrhoea and bronchitis alone claimed a large toll of young lives. The high infant mortality rate — up to one in four in some places — was due, believed at least one Medical Officer of Health, to the 'uncleanly habits, improper feeding and a general disregard to the elementary laws of health' of the mothers. But this is an oversimplification. The high infant mortality rate was due just as much to factors over which the miners and their wives had little control. In closely packed insanitary colliery terraces zymotic diseases could quickly assume epidemic proportions. A study made of infant mortality in the Marley Hill district of Durham between 1896 and 1905 made clear the importance of general environmental conditions. It showed that over most of Marley Hill the infant mortality rate of 147 per 1,000 births was exactly the same as the figure for the whole of England and Wales. In the local back-to-back houses, on the other hand, the rate was 221, a full 50 per cent above the local and national average.[111]

The rigours of life in any nineteenth-century mining community fell most heavily on the youngest, those who had had least opportunity to build up resistance to the infections caused by overcrowding and bad sanitation, conditions which the most caring and conscientious of parents could do little to alleviate. Many of course were reluctant even to make the effort. But in the coalfields, as elsewhere, parental indifference was giving way to care and concern, however ineffectively expressed. There 'were many omens and superstitions' to protect babies and in Fife,

> Miners would gladly "lie on" for the father of the invalid child, an expression to mean he would do his own day's work and then return to the pit to do a day's work for the

father in order that he could sit up with the patient all the night . . . The main source of information was the village well nearest the house of the patient. At such a time the "neebors" were always needing water, usually attended to by boys or girls, but on such an occasion the women went themselves for it was there the bulletins were sent out.[112]

The treatment of those children who survived the rigours of early life was determined by three things: shortage of space, shortage of money and shortage of time. The lack of space was crucial. As a girl brought up in the Durham coal-field in the early years of this century points out:

> Families were large, houses were small, so we had to come in only at meal times. Even in the cold weather we spent most hours out of doors . . . We would play on summer days from dinner time until bed time without going home for any food. The taps were in the street so it was easy enough to put ones mouth under and get all the drinks one needed.[113]

Working-class children were expected to make their own amusement, preferably outside. All over the country they played marbles; made slides in winter; built boats ('Many a boat the writer has sailed, when a boy, on one of the large lakes in the centre of the colliery village, after the rains.'); played schools ('I spent my childhood teaching the walls in the backyard. One big bump in the very uneven walls was the big dunce in my class at school.'); played hand-ball and whip and top; fought with 'slings' (balls of rolled up paper at the end of three or four yards of string); played 'tip cat' (hitting a piece of wood into the air with sticks) and so on.[114] Popular too were games based on death, death which they saw so often when a baby sickened or when a miner was injured in the pit.

> Our pleasure was picking the flowers, and playing the "lovely" game of funerals. One child would lie on the grass full length with arms duly crossed on the chest, and the other children would heap grass and flowers on top of him.[115]

With young children religion was not terribly popular. 'Religion in those days was a scarey thing. We believed that for every lie we told we would pay for in the here after by having a hot poker pushed down our throats.'[116] A South Wales miner remembers that at his local nonconformist chapel there were 'lurid descriptions of hell where the flames were always rolling at terrific heat . . . and I was terribly frightened at the thought of ever getting into that fire.'[117] But older children often found a new interest in the chapel. At Tow Law in the south-west of the Durham coalfield just before World War I

> There was a zest for the Christian religion, but concealed was an unconscious mixed motive in the mingling of the sexes . . . The Churches, especially nonconformist churches, were sources of social activity. It was here that boy met girl, and parents were relieved that this association of the sexes was taking place in the best of all places.[118]

The medical attention given to school-age children was no better than that received by their younger brothers and sisters – or, for that matter, than that afforded to other working-class children.[119] But at least, in the words of one of the Children's Employment Commissioners, 'Those who survive are strong, because the weak soon perish.'[120] Many infectious diseases were almost ignored. Parents in the South Yorkshire coalfield regarded the 1883 whooping cough epidemic 'with indifference' and at Hemsworth healthy children were allowed to play with their friends who were suffering from scarlet fever.[121] On a Saturday afternoon at the end of the century the four year old son of a Darlaston miner fell into a bucket of scalding water which his mother was going to use to clean the back kitchen floor. Yet it was not until the Monday afternoon, two days later, that his parents finally sent for a doctor.[122]

Shortage of money – or at least the nagging uncertainty of never knowing how much the next pay would consist of – meant that, as in all working-class families, clothes were handed down from one child to the next. 'Certain things cost half a shift. Others a whole shift. Some even more than that. Boots were one of the big items that had to be "saved" for.'[123] In times of the very worst depression even this system could break down. A 'house-to-house visitation' of Merthyr Tydfil during the bad years at the end of the 1870s showed 'a fearful state of things – . . . the children in utter rags'.[124] Many miners' children probably never knew when their birthdays were. In a large family they were too expensive: 'No birthdays were celebrated, but then "there were too many".'[125]

The abiding impression of child care in nineteenth-century mining families – indeed in working-class families as a whole – is of a harassed mother doing her best to cope with too little money in a small, insanitary home.[126] She probably tried hard to be fair. 'Life seemed to be made up of turns,' remembers a collier's daughter,

> Who would sit nearest the fire, who would get the crusts off the loaf, who would get the legs of the rabbit . . . Who filled the boiler, who filled the coals, who slept in the middle, the warmest place when three in a bed. Who went to the farm, or who went to the butchers. Who scrubbed the netty seat and who did the knives and forks. Who washed and who dried the dishes.[127]

Inevitably mother's patience soon became exhausted: 'Disobey her, or cross her in any way and out came the tawse . . . This was a leather strap with tails cut half way up.'[128]

It is difficult to detect much change in the treatment of miners' children over the course of the nineteenth century. But one way in which it did change, and for the better, was that children seem to have been left alone in the house less often. A study of any early nineteenth-century coalfield newspaper will reveal a comparatively large number of children injured while left in the house alone. In 1823, for instance, a North Staffordshire collier and his wife went to a pub in Burslem in order to collect his wages, leaving behind three children aged twelve, six and two. The couple stayed drinking later than they had intended and when they returned home they found the children burnt to death.[129] Tragedies like this became less common as the century progressed. Perhaps as families became smaller and more settled it became easier to arrange for a neighbour or friend to keep an eye on the children. Another way in which it is possible that child care improved was that as families became a little smaller and a little better off, parents found it possible to begin to devote more time to their children. A boy brought up near Washington in the high-wage Durham coalfield remembers that at the beginning of this century families had sing-songs together and 'Parents would spend time to listen to their children repeating their multiplication tables, counting on a bead counter, etc.'[130]

Naturally it is hard to generalise for there were good parents as well as bad. But there seems to have been little change in the way in which most miners' children were brought up. Most mothers appear to have tried their best for their children in most inauspicious circumstances. The first mines inspector, H. S. Tremenheere, was not always the most reliable or sensitive observer of the mining scene. But his evaluation of child care in the mid-century Black Country accords well with the evidence available from other sources. Though most wives did not work (at least by the end of the century), Tremenheere's views are worthy of more general application.

> The mother is often absent at work all day, or, if not, is as little capable of training up her children under a steady, temperate, and intelligent discipline, as those with whom she leaves them. Inordinate indulgence is almost the only mode she seems to possess of testifying affection, and this often alternates with anger and violence.[131]

Few working-class parents set much score by their children's education and in this the miners were no exception.[132] The early nineteenth-century 'dame' schools provided little more than a rudimentary child-minding service while there were obvious objections to the elementary schools opened by some colliery owners round the middle of the century.[133] An Ebbw Vale miner protested against the teachers in his local schools

> who spare no pains in trying to impress upon their docile minds the first rudiments of education, and the necessity of paying due homage to those who have enriched themselves at the expense of the workmen's capital.[134]

Nor was the introduction of state education after 1870 any more successful in reconciling the mining community to the advantages of formal schooling. In both voluntary and board-schools children continued to be taught in large classes; in a Rhondda school in 1913 one certificated teacher was having to take classes of 105 and 106 respectively.[135] In such circumstances it is not surprising that teaching tended to be mechanical and discipline harsh. 'School was a drab, grey routine of unexciting tasks.' 'School was very monotonous under the Londonderries,' remembers the daughter of a Durham miner.

> Religion reading, writing, Arithmetic – home time and homework . . . We only learned patriotic songs when some V.I.P. was expected, but on a Friday afternoon after play-time the boys would have drawing and the girls needle-work. Even so this was dreary for the boys would draw a jam jar and the girls would stitch a hem on a piece of calico about 5" x 3", the object being to get even stitches.[136]

It was strict as well as boring. 'Too often,' according to a South Wales miner, 'the boy who was a bit slow at answering questions, dull at arithmetic, or hopeless at grammar, was severely and wrongly Caned on his hands as well as on the part where he sits down.'[137] In Durham too 'School was a terrible place. It was the rule of the rod.'[138] 'In Those Days, Some Teachers Enjoyed Hitting The Children For Nothing. The Mothers Were Often Down At The Porch Playing 'Ell' At Them.'[139]

Despite the occasional education success story, the mass of oral and autobiographical evidence is too insistent – and consistent – to be ignored or dismissed.

> We boys were all looking forward to the day of our emancipation. Our keenest desire was for freedom; freedom from the boredom and imprisonment provided by the four walls of a cell, called school.[140]

Few parents countered this desire very forcibly. Even when elementary education was finally made free in 1891, continuation into secondary education meant extra expense and the loss of an additional income. After Standard VI, remarks a mineworker from Fordell in Fife, the next stage was Dunfermline High School. The expenses caused by sending a pupil to Dunfermline, providing school books (as this was long before the days of "free" books), was entirely beyond the means of the Fordell miner.[141]

Education did cost the family money; teaching methods were often dispiriting; and discipline was frequently harsh. But whatever may have been said, these were not always the real reasons for the miners' lack of commitment to their children's education. The essential explanation, as the first mines' inspector was quick to appreciate, arose

from their low estimate of what is meant by education, and of its ultimate value to their children . . . the standard aimed at by the parents for their children being only a little writing, a very little ciphering, and the power of "reading", however imperfectly.[142]

Not surprisingly many miners and their wives believed that they could best serve their children's interests, as well as their own, by encouraging their sons and daughters to prepare themselves for entry into the working-class domestic and labour market. Girls were encouraged to make themselves useful around the house. 'Mother wasn't keen on reading "trash". All books were "trash". She thought one's time was better spent on mending, darning, knitting etc.'[143] Boys too were expected to do odd jobs like digging the allotment and bringing in the house coal from where it had been dumped.

Periodically on leaving for morning school, the coal-leader could be seen, with two horses and carts, one following the other, delivering not mere sackfuls but whole loads of coal, twelve to fifteen cwts., neatly deposited at the back of almost every house in the row, close to some self-erected shed though it was never so named but rather called the coal-house or coal-cree.

By mid-day, here and there, a miner's wife would be seen bravely struggling to cast some of the coal into the cree. Some boys, usually in pairs, with an eye to earning extra money would knock at the doors and shout 'Schull (shovel) your coal in, Missus,' and for two or three coppers get it quickly into the cree.[144]

In some parts of the country, in South Wales and the West Midlands for example, it was the boy's job to take his father's dinner to him while he was at work. The pit yard was full of temptations. In 1863 a six year old boy was taking food to his father in a pit at Blaenavon near Pontypool when he jumped onto the hitching between two wagons, only to be knocked off and killed when the engine started.[145] Most parents hoped that while still at school their children would be able to make some sort of regular contribution to the family budget. The son of a Derbyshire miner remembers running errands for the landlord. But his

> first real earnings came ... as a newsboy selling evening papers after school hours. I could earn two or three shillings a week by these means. The date was 1897 and a shilling would buy a stone of flour (my mother baked her own bread like most housewives in mining villages) or a pound of butter or three pounds of lard. Half a crown would go a long way towards a pair of boots or trousers.[146]

A girl brought up in the South Yorkshire village of Denaby started work when she was about twelve. 'I did potato churn-ing ... for a fish and chip shop. You used to do two peggy tubs full of potatoes for threepence. Four churns to a tub, hand churned ... And I used to go on a Saturday and help to scrub out.'[147]

Children were encouraged to go out to full-time work. All over the country, it was the custom for children living at home to hand over their wages to their parents. And as Ned Cowen points out, the parents kept most of it.

> My first job was on the pit surface cleaning out tubs. My first pay was nine shillings and twopence less threepence offtakes. I handed my pay to my mother and was given threepence pocket money. A halfpenny was spent on broken candy and a halfpenny on hot peas from "Granny Barne's" Shop. On the Saturday night I went to Gaff's Theatre in Paddy's Market at Birtley and for the price of a penny seat in the "Gods" saw "Murder in the Red Barn".[148]

It was never very easy to get jobs for daughters. In the more prosperous coalfields it was quite unusual for girls to hold down jobs outside the home. So in Cudworth, Darfield and Royston in South Yorkshire in 1911, almost 90 per cent of the women and girls over the age of ten were unemployed.[149] Those who did work had as little choice as their mothers. In rural areas there were the farms and in parts of Lancashire, Fife and Yorkshire there was mill work. Otherwise it was domestic service (which might well mean moving away from home), dressmaking (which later gave way to shop-work), perhaps teaching and, at the very bottom of the pile, surface work at the pit.[150]

The range of employment opportunities for boys was little greater. Even in districts where mining was not dominant, the pit was often the only real choice. It was easy to get a job in a rapidly expanding industry; the job was familiar; there was probably little travel; the pay was relatively high and the formalities were minimal. At the Babbington (Cinderhill) collieries in Nottinghamshire it was usual for the manager to interview men whose sons were about to leave school, saying for example, 'Tell your Johnny to start ganging at No. 4 Pit on Monday morning.'[151] This does not mean though that parents were blind to the dangers. How could they be? When George Parkinson's father left him on his first shift underground in 1837, he had 'tears in his eyes' and said to himself, 'Aw wish ye'd byeth been lasses.'[152] A boy brought up on the edge of the South Staffordshire coalfield at the end of the century remembers that his father, who had been crippled in the pit, was insistent that his son should not follow him underground. He searched and searched for work but finally ended up at the local colliery – where his father had received his injuries.[153]

It is ironic that the high wages that the young miner could earn were a factor both in persuading parents to allow their children underground and in causing so much domestic tension. Miners enjoyed their maximum earning power when they were in their late 'teens and early 'twenties; thereafter their wages tended to decline. So it is not surprising that sons should match and then overtake their fathers' status and earning capacity as the latter slowed down, were relegated to oncost work or even put out to graze on the surface. It was

not a situation likely to foster domestic harmony.[154] Never-
theless it does seem that mining families tended to be particularly close-knit. Many children stayed at home so long as
they were single; they often made their first home with
their parents and generally maintained close ties with their
families for years afterwards. Insofar as it is possible to
generalise, it appears that grown-up sons and daughters did
their best to look after their ageing parents. The trade unionist,
P. M. Brophy, claimed in the 1840s to know of daughters
around Newcastle who were going into prostitution in order
to keep their nearest and dearest out of the dreaded workhouse.[155] Few went to these lengths but those able to remember mining communities before World War I insist that 'Old
people were never neglected. They always seemed to be the
responsibility of their own family.'[156]

> It was a code of honour strictly adhered to to show respect
> for old people . . . and everybody would help old people to
> get from A to B, or run messages or fill pails of water or
> coals. We thought nothing of it, but that it was the done
> thing . . .
>   There was more socialism practised in those days than
> there is now.[157]

It is easy to be cynical about such claims. Yet what little
research there is on nineteenth-century working-class families
does suggest that shared hardships, the inadequacy of the
welfare services and a recognition of parental efforts and
sacrifices tended to bind families together.[158] Certainly mining
families 'appear to have possessed a concept of familial support which went beyond a mere reciprocation of duties and
services or occasional help in times of crisis'.[159] It was the
family which helped its members to survive the rigours of life
in the coalfields. The nineteenth-century mining family was
strong and resilient and altogether more responsible than it
is given credit for.

6

The Miner at Play

The nineteenth-century miner enjoyed a reputation for hard drinking even more formidable than that attributed to other working people. The coalfields rang to denunciations of his drunkenness, with the clergy particularly vocal in their criticisms. Before 1800, one Shropshire vicar was attacking local miners as drunkards who turned their 'enormous bellies into moving hogsheads'.[1] Others were more specific. The Rev. T. Davies of Pontypool blamed strong drink for the poor attendances at his services. On any Sunday evening, he said in 1846, almost half the adult population of the town was to be found in one or other of the local pubs and beer shops.[2] The colliery owners blamed drink for causing laziness and absenteeism. 'I never yet knew high wages obtainable,' wrote one manager, 'but drunkenness, idleness, and loss of time were sure to be the consequences, as young men whose wages exceed their reasonable wants of food, lodging, and clothing, generally spend the surplus in the Ale Houses . . .'[3] The miners themselves knew only too well the results of heavy drinking. 'It is indeed a pitiable sight', remarked a Durham man in 1863, 'to see both old and young — father and son — staggering home, up and down the public streets, scarcely able to walk at all, and not knowing what they are doing.'[4] Only the licensed trade was happy. When the Wheat Sheaf Inn at Bilston in the Black Country was advertised for sale in 1818, much was made of its 'excellent Situation, being surrounded by extensive Coal and Iron Works'.[5] Nor have modern social and labour historians seen any need to challenge the miners' reputation for heavy drinking. Dr Challinor identifies drinking as the most popular pastime of the Lancashire and Cheshire miners while Dr Burgess feels able to talk

142

about the 'habitual drunkenness prevailing in many mining villages'.[6]

There is certainly a good deal of evidence in support of the view that the miner liked nothing better than a drink. The local newspapers published in every coalfield were full of stories of drunken, rowdy, disorderly and violent mineworkers. Here are just a handful, taken at random from the pages of just one of these papers, the *Wolverhampton Chronicle*. In December 1824, a pit's company of thirty men met for a drink in Tipton; two of them began to quarrel on the way home, one receiving a fatal stab wound in the thigh. In 1838, a miner named Joseph Eades stayed drinking with a friend in a beer shop at Lower Gornal until after midnight; they had an argument, stripped off and started to fight. Twenty years later a drunken collier impersonated a policeman and made a married couple accompany him to the station house where he accused them of breaking his windows.[7] A glance through the files of any nineteenth-century newspaper will confirm that incidents like these were repeated *ad nauseam* wherever miners went to drink.

The belief that miners were particularly prone to heavy drinking has been reinforced too by one feature of colliery communities which was noticed in an earlier chapter, the large number of drinking places which they contained. It will be recalled for example that in 1819, very soon after the opening of its colliery, the Durham village of Hetton-le-Hole boasted one public house for every thirty-three of its inhabitants. Hetton was somewhat exceptional. But even so the ratio of people to pubs in mining areas commonly stood at between 100 and 200:1, much the same as in the country as a whole and equivalent perhaps to one public house for every twenty or forty adult males.[8]

Yet a simple count of the number of pubs will produce a very serious underestimate of the number of outlets at which alcoholic drink could be bought or consumed. Even in rows of houses too small to support their own pub, drink was never far away. The closest pub to the few colliery houses at Bewicke Main, outside Gateshead, was a mile away at Kibblesworth. Accordingly, right up to the end of the century and beyond, a beer cart used to call at Bewicke Main without fail every pay day.[9] In larger settlements there were always many outlets other than the pub. In early nineteenth-century Scotland a licence to sell drink could be bought for as little as 2s. 6d. (13 pence). The result was that spirits were on sale almost everywhere. Of 126 occupied shops in the Coatbridge district of Lanarkshire in 1844, sixty-six — every other one — were selling spirits. The thirsty miner could go to any one of thirty-three specialist spirit dealers, twenty-five licensed grocers, four eating houses, four stores and three inns.[10] Then there were the beershops. The 1830 Beer Act took the tax off beer and allowed any ratepayer to open his house as a beershop upon the payment of two guineas to the excise. Beershops sprang up everywhere; by the end of 1830, 24,000 had been opened. Being out of the public eye, they may well have been more attractive to many workers than the conventional public house.[11] Later in the century yet another type of drinking place came to jostle with the pub and the beershop for the miner's custom. The introduction of Sunday closing to Scotland in 1853 and to Wales in 1881 gave a great fillip to the shebeen or bogus club where 'beer is taken in casks from a wholesale dealer and consumed upon premises which are unlicensed.' By 1889, Cardiff alone had over 480 shebeens, many of them run by workmen and all of them selling adulterated beer, described by one customer as 'a cross between senna and vinegar'.[12] It was not only in Wales and Scotland that the shebeen was able to attract the drinking miner. The Derbyshire pit village of Shireoaks became notorious for its shebeens, despite the fact that 'time after time the police . . . made big hauls, and the justices . . . imposed exemplary fines.'[13] Finally towards the end of the century the miners began to organise their own legal drinking clubs. By 1905 there was a Working Men's Club in virtually every village in the central Yorkshire coalfield, with the largest villages supporting as many as four.[14] One of these, the Wath-on-Dearne Working Men's Club, was remembered by an early steward (who went on to become a Labour cabinet minster) as

> a microcosm of socialism, an institution owned and managed by its members, buying wholesale and selling

retail to themselves at as near cost price as possible, with
profits eliminated . . .
The club itself consisted of a large room with a stage at
one end where vocalists, comedians and other variety
artistes hired by the Club performed on Saturday and
Sunday evenings, and two smaller rooms which were both
bars. The smaller of these served as the equivalent of a
saloon bar.[15]

In the coalfields, as in the whole country, sellers of drink
were as numerous as sellers of food.[16] So the thirsty miner
never had to go far for a drink. And he never had to wait
long either. During the first half of the century beerhouses
were allowed to stay open from four or five in the morning
until ten or eleven at night; while on Sundays they opened
from one to three in the afternoon and from five till ten in
the evening. Public houses closed only during the Sunday
morning church service.[17] Small wonder then that a scan-
dalised Tremenheere should complain in 1844 that it seemed
in parts of the Lanarkshire coalfield as though there was no
control at all over drinking hours.[18] Even at the beginning of
this century, pubs, beerhouses and clubs everywhere stayed
open from six in the morning until ten or eleven in the evening
while on Sundays they opened for two hours at lunchtime
and for three to five hours in the evening.[19]

Yet it was not simply easy access to alcohol which encouraged the miner to drink. The motivation was far more
deep-rooted. For much of the period the very nature of pit
work, the way in which he was paid and the home in which
he had to live all tended to lead the miner into the taproom.
Mining was hard, dirty and dusty. What could be more welcome at the end of an exhausting shift than a long refreshing
drink? It was an especially attractive proposition in the early
nineteenth century when drinking water was scarce and un-
safe, when tea and coffee were expensive, when even fresh
milk was dangerous and when intoxicating drink was believed
to impart both stamina and physical strength.[20] Nor of course
did the early nineteenth-century custom of paying wages in
the pub or in the form of drink do anything to encourage
sobriety. Indeed the irregularity of earnings, which was always

so marked a feature of mining life, probably did a good deal to stimulate overindulgence. It was only too easy to regard the money earned in a good week as a bonus, a surplus to be spent and enjoyed immediately and every increase in take-home pay was followed immediately by an increase in the amount of alcohol consumed.[21] The attractions of even the seediest pub, club or beershop were magnified too by comparison with the squalid, unhealthy, overcrowded homes in which so many miners lived. It is not surprising that Sunday closing was so unpopular when it was introduced to Wales and Scotland.

> How would these very good people [the proponents of Sunday closing] like to live days, weeks and months underground without a sight of the sun, and then on a wet Sunday to keep within doors all the sunless hours, except while attending divine worship? Oh, these very generous people have their nice cosy clubs or homes which they enjoy every day. But the collier has to live in discomfort in a small home, and for near six months in every year never sees the sun, except on the first day of the week.[22]

The pub performed a vital role for the whole of the working class. 'The price of a drink was their entry-fee to comforts which the prevailing social situation enabled them to enjoy only communally.'[23] The pub was the place to meet friends, hear the news, denounce the management and talk about work — 'more coal is still filled by miners in the pubs than they ever fill while they are at the pit . . .'[24] Then the pub was the home of all sorts of games and contests, many of them long since forgotten. Linnet-singing matches were held in Lancashire. The miner, or other contestant, paid a one-shilling (5 pence) entrance fee, received sixpence (3 pence) worth of ale and could compete for prizes like a copper kettle. Pastry feasts were specially popular round Blackrod and Orrell. Courting couples paraded in front of the drinkers; the prettiest girl was given a pint of rum and a sixpenny pastry while the ugliest received the booby prize of a black pudding.[25] Every single pub offered a range of attractions. Take the Angel Inn between Bowers Row and Allerton Bywater, about three miles from Castleford.

In the summer, miners would play marbles and quoits out in the yard at the back of the pub, their beer being brought out to them, or fetched by each in turn . . . They had dominoes teams too . . . Cards were always being played . . . There was also another game that does not sound very posh. This was known as the spitting game, and was always played in the tap room . . . The men were constantly chewing tobacco, so they were always spitting, and they became real experts at it. They used to bet on how far they could move away from the spittoon and land their spit dead centre . . . Small dart leagues were also formed.[26]

There were obviously enormous pressures on the mine-worker to meet and drink in the local beershop or public house. And there seems no doubt at all that most miners did enjoy having a drink with their friends. They drank heavily on pay-night and on special occasions. Fairs and wakes and, in Northumberland and Durham, the annual binding, were all early nineteenth-century excuses for overindulgence. The Houghton feast in Durham was said to be the scene 'of drinking, gambling and fighting' while it was claimed in Wolverhampton in 1824 that the 'almost weekly occurrence of Wakes, Horse and Donkey Races, &c. in our immediate vicinity [are] invariably the resort of the idle and disorderly, and the scenes of gaming, riot and drunkenness . . .'[27] Family celebrations also saw a good deal of drinking during the first half of the century. Children christened in the North-East of England were accompanied to the church by family friends who then moved on to the local pub.[28] Weddings too were an excuse for a good time. At a wedding at Chesterfield in 1841 a party of thirty-seven drank '2 gallon of gin, 2 gallon rum, and a load of malt brewed for the occasion'.[29] By the end of such celebrations:

The barrel's found no more's to broach.
There' but a pipe for everyone;
The dear tobacco's almost gone.
The candles in their sockets wink,
Now sweat, now drop, then die and stink.
Intoxicating fumes arise.

They reel, and rub their drowsy eyes.
Dead drunk, some tumble on the floor,
And swim in what they'd drunk before.
'Hiccup,' cries one. 'Reach me your hand.
The house turns round. I cannot stand.'
So now the drunken senseless crew
Break pipes, spill drink, piss, shit and spew.
The sleepy hens now mount their balk,
Ducks quack, flap wings, and homewards walk.
The labouring peasant, weary grown,
Embraces night and trudges home.[30]

But pay night was the big one. As many as 10,000 people
used to flock into Airdrie on pay nights in the 1840s. 'Scenes
of uncontrolled license ensue, which [with only one super-
intendent, four constables and four or five part-time assis-
tants] there are no means of either preventing or punishing.'[31]
'Men were always drunk at weekends,' remembers a miner's
daughter brought up in a village where weekly pays were the
rule.[32] Even in villages with a reputation for being quiet and
respectable, there was always fighting and drinking at the
weekend.[33]

The Baff Week is o'er — no repining —
Pay-Saturday's swift on the wing;
At length the blithe morning comes shining,
When kelter makes colliers sing.
'Tis spring, and the weather is cheery,
The birds whistle sweet on the spray;
Now coal-working lads, trim and airy,
To Newcastle town hie away . . .

The young men, full blithesome and jolly,
March forward, all decently clad;
Some lilting up Cat-and-Dry Dolly,
Some singing The Bonny Pit Lad.
The pranks that were played at last binding
Engage some in humorous chat.
Some halt by the wayside on finding
Primroses to place in their hat.

Bob Cranky, Jack Hogg and Dick Marley,
Bill Hewitt, Luke Carr and Tom Brown,
In one jolly squad set off early
From Benwell to Newcastle Town.
Such hewers as they (none need count it)
Ne'er handled a shovel or pick.
In high or low seam they could suit it,
In regions next door to Old Nick . . .

At length in Newcastle they centre,
In Hardy's, a place much renowed,
The jovial company enter
Where stores of good liquor abound.
As quick as the servants could fill it
— Till emptied were quarts half a score —
With heart-burning thirst down they swill it,
And thump on the table for more.

With boozing and laughing and smoking,
The time slippeth swiftly away;
And while they are ranting and joking,
The church clock proclaims it mid-day.[34]

Few men could find the money to carry on like this through-out the whole week. 'Drink they certainly do . . .' admitted a visitor to the Fife coalfield, 'but there are circumstances which order their drinking and confine excessive drinking to stated times, to wit, the pay night, the day, or it may be the two days following, and holidays, especially at the New Year.'[35] But when miners drank, they drank openly, often heavily, and frequently it seemed, 'to get drunk'. Even occasional drinkers like Arthur Horner's grandfather could end up roaring drunk. He would stay sober for up to six months at a time, then get drunk and end up having a fight in the middle of Merthyr Tydfil — and, need it be said, adding a little more to the miners' already powerful reputation for heavy and sustained drinking.[36] Moreover working men did not have 'the opportunities that the wealthy have of concealing their vices; if they are drunkards . . ., they can seldom hide their intemperance . . . from the public gaze; in this way a few drunken brawlers often bring disrepute upon the whole body of workmen to which they belong.'[37]

There is a good deal of special pleading here; but there is also a great deal of truth. Indeed the strength of the miners' reputation for both heavy drinking and for religious nonconformity should warn against any easy condemnation of coalfield drinking habits. Overindulgence was not the universal custom and it is wrong to assume that going for a drink was necessarily incompatible with consideration for the family or with an interest in thrift and self-help. For most miners a long session at the pub or beershop was a pleasure to be enjoyed only when they had been paid or when there was a special occasion to celebrate. While there were many miners who drank to excess, a careful study of the available evidence suggests that there were also a good number who drank little if at all.[38]

So it was that the miners acquired their particular notoriety for overindulgence. And this reputation lived on long after it had ceased to have any firm basis in fact, if indeed it ever had done. For there is not the slightest doubt that the scale of miners' drinking declined as the century advanced. This was part of a broader, national trend for, despite appearances, the incidence of drunkenness everywhere did begin to diminish from about 1830 onwards.[39] Weddings and christenings in the coalfields were becoming much more sober and respectable. The geordie pitman-poet, Thomas Wilson, could look back in 1843 at the times when 'the pitman's child was taken to the church in the forenoon, accompanied by a large party of friends, who returned (after "getting up the steam" a little at the public house nearest the church-gates) to a hot and substantial dinner, followed by both tea and supper: — now, this ceremony is generally deferred until the afternoon, and is attended by the sponsors only, who return to — "a cup of tea".'[40] But memories of earlier excesses lingered on.

> When some of his friends in Lochee [outside Dundee] had got to learn he was going to a miner's wedding in Fife he was advised to prepare himself for the occasion which would certainly arise, and he produced from a coat pocket some sticking plaster and a bottle of linament which, he was told, he would find a use for. He would take back these articles and would have the pleasure of saying that they were not required.[41]

One reason for the gradual decline of intemperance lay in the development of the smaller and more settled family unit which was considered in the previous chapter. At the same time slowly improving standards of accommodation, improved material comforts, better water supplies and the greater availability of tea, milk and other non-intoxicating drinks combined to make the beerhouse and public house less attractive.[42] Yet another reason for the decline in drinking was that the employers came to believe that the curbing of drunkenness was essential if they were ever to acquire a steady and reliable workforce. Slowly the legislation of 1842 prohibiting the use of the pub as a pay office and the payment of wages in the form of drink began to take effect. Dying out too by the end of the century was the custom of providing lavish feasts and celebrations whenever a pit was opened or an employer's son came of age.[43] Some employers, it is true, ran pubs purely for their own gain.[44] For some, however, the drive for profit was tempered by the desire to alter their employees' drinking habits. Sobriety was sometimes felt to be more at risk from spirits than from beer; accordingly in the middle years of the century at least one Scottish coalowner forbade the sale of spirits – though not of beer – anywhere on his land.[45] The Fife Coal and Iron Company ran its own pubs but 'with a view', it was claimed, 'to furthering order and sobriety among their workmen'.[46] A few employers, like the Durham firm of Pease and Partners, actively encouraged temperance work, sacked habitual drunkards and opposed the licensing of new premises.[47]

A more common strategy was to attack the drink problem by opening non-licensed clubs, tea-rooms, institutes, libraries and reading-rooms as counter-attractions to the iniquitous public house. A number of employers, including such leading North-Eastern coalowners as Love and Pease, the Bearpark Coal Company, Lord Londonderry and the Earl of Durham, helped to start British Workmen's Public Houses, pubs which were to offer 'all the advantages of drink taverns without the drink'.[48] Often colliery-owned clubs and institutes were not allowed to sell alcohol. The Public Library and Moray Institute, which was built in the Fife mining village of Kelty at the turn of the century, housed a circulating library, reading, recreation and billiard rooms but it sold no intoxicating drinks and was deliberately sited some way from the nearest pub.[49]

There was a much longer tradition of providing libraries and reading rooms. These were doubly attractive from the employers' point of view. Although relatively inexpensive to set up, they offered the hope at least of modifying the miners' outlook on the world. The large firms of Lanarkshire and the North-East of England were the first to enter the field. Already by the middle of the century there were libraries in Scotland at the Govan colliery and at the Clyde and Monkland ironworks. The library at Govan charged working colliers sixpence (3 pence) a week, was 'open every evening; . . . is furnished with a long reading table, and is well lighted and warmed'.[50] But the reading rooms and lending libraries which had been opened in Northumberland and Durham after the strikes of 1839 and 1844 did not last long. By 1850 'they were very little used; and have in some cases been abandoned.'[51] The great period of library and reading room foundation came in the second half of the century. Old buildings were reopened and new ones started wherever large scale mining took place. They sprang up in South Wales, Yorkshire, North Staffordshire, across Scotland and throughout Northumberland and Durham. From 1850 onwards, the Seghill (Northumberland) reading room was open for two nights a week providing its members with a choice of 700 to 800 books and a selection of London newspapers and magazines. At the Shelton colliery and iron works in North Staffordshire Lord Grenville heated, lit and furnished two rooms in the colliery yard to be used as a reading room. Three newspapers were provided and strict rules enforced. One room was designated for quiet reading, while the other was set aside for conversation and reading aloud.[52]

These reading rooms and libraries were started by the employers as part of their attack on what they saw as the miners' obdurate intemperance and irresponsibility, as part indeed of a much larger entrepreneurial offensive designed to turn the workers into a reliable factor of production.[53] At first sight the initiative appears doomed to failure. The reading rooms and libraries were often poorly constructed (the room at Seven Sisters in the Rhondda was built of corrugated galvanized sheet); badly sited (readers at East Cramlington in Northumberland were regularly affected by the nearby gas works); and full of books like *Abbott's Young Christian*, the *Backwoods of Canada*, and the *Rise and Progress of Religion in the Soul*.[54] Yet what is so striking is the apparent enthusiasm with which a minority of colliery workers parted with their money in order to have access to them. In the middle of the century twenty-three 'average' pitmen belonged to the Seaton Delaval colliery library — only two or three per cent of all the men employed.[55] An eighth of the 4,000 or so workmen at the Monkland Iron Works paid a penny a week to read the magazines and newpapers in the reading room or to borrow from among the thousand volumes housed in the attached lending library.[56] And despite complaints of poor support for the Dudley Colliery Institute and Reading Room in Northumberland, it transpired that in the early 1860s it was being used by about a third of the men employed at the pit.[57]

The more serious men took to the libraries and reading rooms immediately as a welcome addition to their meagre educational and recreational facilities. It needed longer for them to gain more general acceptance. But the reading room's descendant, the Miners' Institute and Library, was to become very influential in South Wales at the end of the century. By about 1910 almost every large mining village and town in the coalfield had built, or was building, its own institute. Only in the Rhondda did they manage to escape the dead hand of the local minister or colliery owner, yet everywhere they played a powerful role in community life. When free elementary education was eventually introduced, the colliery workers diverted the poundage which had been deducted for their children's education towards the support of the local Miners' Institute. The Maerdy Library was one of those maintained by the men. When a columnist from the *Glamorgan Free Press* visited it unexpectedly in 1892, he found that

> The room was crowded. On all sides could be seen the hard-worked colliers, with the characteristic rim of coal dust on the eyelids. A glance round the room showed

me immediately the interest and earnestness with which each was devouring the contents of some book or newspaper . . .[58]

Reading rooms, institutes and libraries were the most important adult educational self-help institutions to be found in the coalfields. But they were not the only ones. A few miners went to mechanics' institutes and later in the century a good number presumably joined the new, free libraries.[59] Indeed one Lancashire collier was accused of being a socialist by the trade union leader Sam Woods because he wanted Wigan Public Library to open on Sundays.[60] Night schools, mutual improvement classes, literary and debating societies, the Workers' Educational Association and the University Extension Movement also made some limited impact towards the end of the century. Yet, as one South Staffordshire clergyman put it, 'great big lads who cannot read more than ABC soon get disheartened.'[61] The only forms of adult education to attain genuine popularity in the coalfields were first aid classes (often organised by the St John's Ambulance Association) and public lectures on some aspect of mining. The 1880s and 1890s saw first aid classes opened throughout the colliery districts, a development which gained royal approval in 1893, when at Windsor Great Park, Queen Victoria received 400-500 miners who had been trained in first aid at the Derbyshire and Nottinghamshire collieries belonging to Colonel Seely.[62] Mining lectures sometimes attracted huge audiences. On one occasion at the end of the century, a mining lecturer addressed a Rhondda audience of over 900 people and at the very end of the period, a lecturer in mining from Manchester University was giving talks to hundreds of colliery firemen at Wigan and at nearby Ashton-in-Makerfield.[63]

Temperance in colliery communities was encouraged therefore by slowly improving domestic circumstances and by the concern of the owners to create a sober working-class that would be able to support itself in every eventuality. It was encouraged as well by the decline of popular recreations like animal-baiting and animal-fighting which had always been closely linked with wakes and feasts, large crowds and alcoholic excess. At the beginning of the nineteenth century cockfighting was practised in many districts, with bull-baiting specially popular in the Derbyshire, Shropshire and Staffordshire coalfields.

> The numerous collieries of Tipton and its vicinity, have produced thousands of beings, who being trained in their infancy to the love and practice of Bull-baiting, possess nothing human, (when arrived at man's estate) but the form. Ignorant, vulgar, and wicked to excess, their ferocious rage for this bloody and barbarous amusement knew no bounds, would brook no control.[64]

The first three decades of the century saw a vigorous campaign against popular blood sports. Bull-baiting was forbidden at some wakes and the Earl of Dartmouth introduced gymnastics to a West Bromwich wake 'with the praiseworthy intention of putting a stop to the brutal system of bull-baiting, and other cruel practices, which have hitherto been characteristic of the wakes in the colliery and ironworks districts'.[65] The South Staffordshire Association for the Suppression of Bull-baiting was formed in 1824, the baiting of animals was made illegal in 1835 and by 1840 the practice had been almost eliminated. Badger-baiting, dog-fighting and cockfighting lingered on in rural areas for another half century or so: in the Lothian coalfield, for example, cockfighting went on in secret well into the 1880s. In the main, though, they lingered as occasional brutalities rather then as regular pastimes.[66]

The reduced recreational opportunities resulting from the disappearance of these blood sports were more than made up for by the emergence of new interests and possibilities. In coal-mining as in other industries, the dominance of drink was being undermined not only by higher earnings and new domestic comforts but by improved communications and the emergence of new, mass leisure pursuits. It was becoming feasible to spend money on drink as well as on other recreational activities. The improvement in communications was important. The railways quickly penetrated the coalfields, bringing with them cheap Sunday fares and special excursion trips. There was a proliferation of proprietorial treats. In September 1858, for example, the owners of the Wombwell Main col-

liery near Barnsley chartered a twenty-four carriage train
to take nearly 700 workmen and their families on a trip to
Thorne on the far side of Doncaster.[67] The national rail
network was almost complete by the 1880s; in 1905 the
third class return fare from Barnsley to Blackpool was 3/3
(16 pence), still more than enough for a surface worker to
find but only about a third of what the best local hewers
could make in a single shift.[68] It is easy to exaggerate the
impact of the railways. 'The railways', claims the historian of
the Derbyshire miners, '. . . were like a magic carpet spiriting
the miners and their families away from their drab surround-
ings to the El Dorado of the seaside resorts.'[69] In reality life
was neighbourhood-centred and horizons remained limited;
as one Welsh hewer pointed out, even 'In the year 1890 the
prospect of a visit to London was as distant and impossible
to a South Wales miner, as a visit to Venice or Moscow might
be today [1951]. It was possible only to the abstemious
single man.'[70] But gradually the railway, the bicycle and the
horse and cart did all help to break down that isolation which
was to be found at the root of so much heavy drinking.

In 1890 Bowers Row had its first transport system installed
direct from the village to Castleford. An old man called
Stockwell, whose descendants are still living in Kippax,
bought a horse and trap so that he could take families or
small groups to do their shopping in the town. Later, he
bought a waggonette and two horses, and installed a
regular service on Saturdays between Bowers Row and
Castleford. He made two journeys, one on Saturday afternoon, the other on Saturday evening. This enabled people
from the village to go to Castleford to watch a football or
rugby match and others to go in the evening to see a live
show in the Queens Theatre.[71]

Better communications created a wider constituency for
leisure, allowing those interested in gardening, music, the
theatre, sport, the church and so on to operate on a wider
scale than would otherwise have been possible. The nineteenth
century saw a surge of interest in gardening and keeping
animals. To begin with these were very much the minority
interests of those whom Tremenheere dubbed the 'steady
colliers'.[72] But miners living in villages and small towns often

had access to gardens and allotments and it was not long before gardening became very popular. As a hobby it had many advantages. It was cheap (and some employers provided fencing and manure); it got you out of the house (especially during the better weather when the house coal trade was slack); there were prizes to be won; and there were obvious benefits in having a private supply of fresh meat and vegetables.[73] By the end of the century families in every coalfield were keeping pigs and growing vegetables. A Durham miner remembers that around his village there 'were scores of allotments with potatoes and leeks, the miners' favourite crops. There were regular shows for prize leeks throughout the North . . .'[74] There was also a great growth of interest in growing flowers, good for the soul perhaps but not quite so beneficial for the stomach. When the Crook (Durham) Agricultural Society organised its first show in 1862 it concentrated upon pigs; by the end of the decade it was a flower show proper.[75] In Fife mining villages

> Competitors selected the different flowers and vegetables they intended to compete with, and not only did they nurse them with all the different ways of bringing them to their best, but they set a night and day guard on them.
>
> Despite careful vigilance, however, many prized flowers and vegetables disappeared just previous to the show, and a competitor had to be very sure of his convictions before he could tell the committee that a flower or turnip on the stand was his property, but had mysteriously disappeared from his garden a night or two before.[76]

And the miners did not just compete in local shows. In the prosperous summer of 1872, South Wales 'colliers flocked to the Royal Agricultural Society's Show' in Cardiff.[77] Round the turn of the century, Fife miners entered various local competitions but 'the show in Dunfermline or Cowdenbeath was regarded as the decider' while in County Durham the Seaham Flower Show as known locally as 'the big flower show . . . That was the day of the year, for people came from all the surrounding villages and there was great fun.'[78]

Gardening of all types achieved wide popularity then, particularly among the older, married men. There was a growing interest too in art and serious music although as might be expected these always remained very definitely minority pursuits. In the early summer of 1880 (a depressed year) between 200 and 300 men from a large colliery outside Merthyr used their days off to visit the Merthyr Art Exhibition.[79] Impromptu singsongs, with or without instrumental accompaniment, were always popular.[80] But there grew up alongside these a strong choral tradition especially in South Wales. Many colliery communities ran their own choral societies. 'Once a year,' remembers a Merthyr miner, 'a local choir, at much expense and labour, devoted a whole week to Opera' with four Covent Garden singers hired to take the leading roles in works like *Il Trovatore, Rigoletto* and *The Bohemian Girl.*[81]

Of wider appeal were the brass bands which seemed to spring up wherever factories were opened and pits were sunk. The instruments were simple to master and relatively cheap to buy, and in many industries the employers sponsored their workmen. In South Staffordshire, the Holly Bank colliery band was allowed to practise in the colliery wagon shop while in Fife, the Fordell band was provided with both the use of a hall and with free paraffin to light it.[82] Nonetheless many bands had to struggle hard. In the autumn of 1862 the Hetton brass band organised a two-day festival with music, refreshment tents, fruit stalls, quoits, shooting galleries and an aunt sally in order to raise money for a suitable band uniform.[83] At Seven Sisters in the Rhondda, members of the band had to pay sixpence (3 pence) a rehearsal to buy music and instruments until by 1906, twenty years after the band's formation, they were able to build their own bandroom at a cost of £90.[84] Brass band music was very popular and very good in colliery districts. As early as 1827 every large colliery in Northumberland and Durham was said to have its own band.[85] In 1869 a group of men from St Hilda Colliery formed a band with the help of a local South Shields musician; within five years they were winning prizes in competition. Some miners went on to achieve international recognition: for example, in the spring of 1912, the Hordern Colliery Brass Band attended the International Musical Contests in Paris.[86]

Of similar broad appeal to mining families were the theatre, and later the cinema. Theatrical performances were staged by the miners themselves and by travelling troups of players. From as early as the 1840s, young colliers from the Monkland Ironworks in Lanarkshire were putting on plays in a local barn and throughout the whole of Great Britain, touring companies were criss-crossing the coalfields.[87] A Durham miner recalls that

> There was at this period a travelling theatre known as "Collett's". It was located often in Sherburn Village just where the four roads meet. The pieces they were going to perform were announced by the orchestra, which consisted of a clarionet played by George, the brother of the proprietor. He was a very strenuous performer, if not musically accurate — so much so that the common remark was that George was trying to blow himself through the instrument. The tragedies he announced were of the most blood-curdling nature. The more horrible the better they were appreciated. And we youngsters left the booth with hair on end and flesh a-creep. The choice pieces were of the "Maria Martin" and "Hallgarth Murder" type. I remember my then, and always, sweetheart and future wife and I were down seeing the "Hallgarth Murder". It was a fine, clear moonlight night, not a cloud to be seen, still and bright. The place was packed, as the theatre was only about a mile from the scene of the actual murder . . . All went well until Clarke had been tried, and was brought out for execution. Everywhere was silent expectation, when just at the crucial point there came a gust of wind, "a mighty rushing wind", which carried away the canvas top of the structure, except certain portions. These cracked like great whips. The women screamed, and there was a universal rush for the doors. Mrs Collett, who had been killed as Mary Ann Westgarth a short time before, stood on the stage wringing her hands and crying.[88]

But the major development was the coming of the cinema. By 1914 it was to be found nearly everywhere and mobile cinemas visited places too small to boast their own picture palace.[89] At Seven Sisters the show was put on in a huge marquee, with probably only half the village children actually paying to get in.[90] Not everybody welcomed the new invention however.

In about 1911 a picture house was built at Kippax, and in 1918 I saw my first moving picture there, called *The Battle of the Somme*. People said my brother Joe was in the film, so we persuaded my father to go with us. At that time the admission charges were 2d., 4d., and 8d.

My father went to the pay box and loudly proclaimed to all and sundry that this was his first visit to such a place. The only reason that he was here was because he had heard that his son, whom he had not seen for two years, was in the film. He said, 'I want three of the best seats,' handed over two shillings, and gave my sister, Amy, and me one ticket each. Then we entered the picture house. The usherette looked at our tickets, and particularly at me, because she was my current sweetheart. She showed us to the back row, whispering to me to sit nearest the end, so that when the picture started she could come and sit next to me. But we had reckoned without my father. He looked at her and said, 'Look here lass, I've paid for the best seats and we are having them. We are going to sit at the front.' With that he strode down towards the screen. The lights were dim and I do not think he even noticed the dark red plush seats we were passing. But when we got to the front we had plenty of choice, for the long wooden benches were nearly empty.

He sat down on the very front row, and of course we had to sit with him. When the picture started it all looked distorted and hideous. He muttered a bit at first, but as usual, he lost his temper and started walking out, shouting, 'Downright robbery, works of the devil,' and other choice words of Christian rebuke.

When we got to the back I turned round and said to him. 'Look, Dad, it's all right from here.' But he would not look; he just said, 'Never look behind, my lad, never look behind!' And then we were outside and our first visit to the pictures was over. He never went to the pictures again and he could never understand why anybody else went, either.[91]

Reading, studying, music, the theatre, the cinema and gardening were all able to attract a number of men away from the beershop and the public house. But there can be no doubt at all that with the single exception of drink it was sport and religion which had the most profound impact on the ways in which mining families used their time and their money. Many of the mining community's traditional sporting (and gambling) amusements maintained their popularity throughout the nineteenth century and beyond. Indoor games like cards and draughts were played regularly. Draughts for instance was popular with early nineteenth-century Scottish miners while at the very end of the period the all-England amateur draughts championship was won by a miner from Nottinghamshire.[92] Gambling was always one of the miners' abiding interests. Many went to the races whenever they could afford it — and often when they could not. Despite the depression of the late 1870s, many Barnsley men considered it 'imperative to go to the [Doncaster] races, seeing they are only a few miles away from the place where they are held'.[93] Infinitely more disruptive of the family economy was the casual, day to day gambling that took place much closer to home. Bets were laid on games of cards and games of marbles.[94] Bets were laid on the handball games that were 'played on the back wall of the end house of any street'.[95] Bets were laid on games of pitch and toss, a game which seemed to exert a baneful fascination. In Yorkshire at the end of the period

> Pitch and toss was also a favourite pastime and seemed a sure way of losing your money. It was played in what they used to call the Pitching School, with point-watchers at all approaches to warn them if any policeman was in the vicinity. They would toss their pennies up in threes, until some were richer and some had lost all the pennies they had.[96]

A Durham worker complained in the 1860s that at the colliery where he worked, 'the gaming system is carried on to a great extent, and its principal supporters are men with families, who are increasing their poverty, neglecting their education, and making no preparation for a rainy day.'[97] Certainly the son of a Durham miner remembers that in the early years of this century his father 'lost his wages three weeks running in the back lane, coming home from the pit'.[98]

There is no way of knowing how typical such incidents were in the different coalfields — or elsewhere for that matter. Traditional pastimes persisted, but they came under increasing challenge from new, often more highly organised and disciplined activities which gave miners and their families (together with other workers) a greater degree of choice and flexibility in their recreational pursuits.[99] Poaching for food gave way slowly to fishing primarily for sport.[100] It was reported from Fife and Clackmannan during the boom year of 1873 that it was not uncommon to see parties of four, hiring a boat on Lochleven for a day's fishing at the rather high price of half-a-crown (13 pence) per hour, and then driving home in triumph in a carriage and pair.[101] Pigeon-racing had always been practised in mining communities but during the second half of the century the sport became more competitive, more respectable and more highly organised. It offered a vital outlet. Fanciers lavished any amount of care, interest and attention on the breeding and racing of their charges. A boy brought up in the Durham pit village of Harden remembers the training of his father's 'milers'.

> These pigeons race but they never travel farther than a mile. They were so trained that they never went up a height . . . they stopped no higher than six feet from the ground, the farther they went up a height, the more time they lost. They were trained on stop watches to race a full mile.[102]

Fighting, football and cricket had also been popular from the very earliest days of the nineteenth century. But like pigeon-racing, these sports too were becoming more structured, more standardised and more popular. In some places fighting gave way — haltingly — to boxing. A West Yorkshire miner remembers that at the end of the century

> There was a stable in one of our local pubs — 'Doddy's' — in which one or two old bare-knuckle fighting miners formed a boxing club. It was there that, for the first time,

we were taught that there was more to boxing than just slogging. They trained us hard, but when they had finished with us we could go to the local feast grounds and challenge the fighters in the boxing booths and, of course, we often did this, with varying results.[103]

Organised football of all types played a vital role in late nineteenth-century working-class communities. Rugby union was destined to become the Welsh national game, but in some Rhondda pit villages the first official matches did not take place until the very end of the century.[104] Rugby League was popular in Lancashire and Yorkshire. But it was the two most popular working-class games, cricket and soccer, that show most clearly the leading role which sport came to assume in mining areas. In the more remote villages few men actually went to league matches but followed their favourite professional clubs at a distance, often having to wait until ten or eleven at night to hear the results.[105] But the local team was also important. A South Yorkshire miner from Denaby remembers that 'During the strike of 1902 Denaby United joined the Midland Counties Football League, and that was a big event in the life of the village.'[106] A West Yorkshire miner remembers that his

village also took great pride in its soccer team . . . We played in blue and white, and had our own song that every man, woman and child used to sing. Chapel people and all others were as one in support of this team . . .
When Bowers played Kippax, our neighbours, we had to have extra police, and gates of over 100 were taken at 3d. each . . . There would be a happy atmosphere if the team won, but if Bowers Row lost, all blinds would be pulled down over the windows, and every villager would be to bed by 8 p.m.[107]

Cricket engendered similar loyalties. During the 1921 strike, a South Wales collier

went across a stretch of waste ground behind our house one morning and saw two grown men amusing themselves by knocking a little ball about with a home-made bat. I joined in, and soon we were half a dozen running about after that old tennis ball. That afternoon the son of the publican offered to sell us a complete set of wickets and a decent bat for five shillings — he had no liking for cricket. We debated this offer for a while, then agreed to start a club — myself as secretary. We accepted the wickets when he agreed to wait until work started for us to pay him. Another man who had been a cricketer and still had his bat was made a life member when he offered to loan this bat. We got more members, and by stinting and collecting amongst the shopkeepers and others we kept up the payments of two shillings a week to pay for pads and balls.

Then we meant to make it a first-class ground. It was a fine display of communal effort: about thirty of us working from morning to night without a grumble carrying stones away, filling up holes, cutting and carrying and laying turf. We appointed a man as groundsman because he claimed to have seen the Oval. He went round with a pail of whitewash and marked the boundary of the outfield. That night — it was moonlight — the soft thud of an axe sounded into the early morning, and next day six small trees that might have been in the way of a good hit, and were inside the whitewash marks, had completely vanished.

A farmer loaned us his roller, and we rolled and trimmed, practised and planned, all day and every day. We could not play on Sundays, so we held general meetings that lasted most of the day, and were held near the cricket pitch so that we could watch the grass growing and stop children and cats walking over it.

After we had started we had matches every day, and the village came to see them. Surely never were a few shillings better spent. The married men among us who had small babies used to bring the babies there while their wives did the housework. If these men were players, and it came their turn to bat, another member of the team would try to pacify the baby while the father did his bit of slogging. Then, when we fielded, the spectators would mind the babies as part of the admission fee.

We crossed over the mountain for one of the matches. We walked; and carried our cricketing gear. More than a hundred and fifty of our supporters followed us the six miles there and six miles back. They saw us win a grand game, watched by — I should imagine — every inhabitant, as well as the four policemen in that other village. We had a concert afterwards, and it was nearly midnight when they let us depart.[108]

By no stretch of the imagination was this a typical incident. But it does give some idea of the impact which locally organised sporting activities could have on a mining community in the late nineteenth and early twentieth centuries. They gave a man the chance to make his mark. Together with higher wages, better transport facilities, new opportunities and changing patterns of family life, they helped to wean the miner from the bottle. He began to garden, study and read; he tried going to the theatre and the cinema and he started to watch or take part in some organised sporting activity.

It was the church however — and especially the chapel — which did most to diminish the power of the pub in the coal-mining districts of Great Britain. By the middle of the century almost every mining community had its own chapel, often very much like the one in the Durham colliery village of New Lambton: 'a square, red-tiled brick building at the end of a long row of miners' cottages . . . Two large windows in front, and a projecting porch covering the doorway between, marked it out as a special building both in structure and purpose.'[109] Indeed it is easy to assume that Methodism was the established church of the coalfields.[110] But it was not without its rivals. Even quite small communities often housed a surprisingly large number of churches and sects. Just before World War I in the Durham pit village of Wingate, 'There was a resident Vicar of the Church of England, two ministers serving the Primitive and Wesleyan, now Methodist Churches; a Bible Christian Church together with an equally virile Salvation Army in their Front-Street Citadel.'[111] Yet Wingate apparently had few members of the miners' most important religious minority, the Roman Catholics. There were large numbers of Catholics in Lancashire and throughout West and Central Scotland.[112] The presence of this important — though often overlooked — Catholic minority did much to weaken the homogeneity of the mining population in these areas. Newspaper reports following the explosion of the Blantyre colliery outside Glasgow in 1877 revealed sharp residential segregation: forty-four dead Catholics coming from 'the Rows'; and forty-seven dead Protestants from 'Dixon's Rows'.[113] On one occasion Keir Hardie introduced Alexander McDonald to an audience of Lanarkshire miners by comparing his work to that of Martin Luther. Two-thirds of the audience were Irish Catholics and the meeting ended in a riot.[114]

It is extraordinarily difficult to make a reliable estimate of the number of miners and their families who belonged to the nonconformist and other religious groups active during the nineteenth century. There are, after all, several different ways in which it is possible to 'belong' to a religious group but the best yardstick is probably that of attendance at worship. Attendance was by no means uniform. Violations of the Lord's Day were 'common and disgraceful'; in many parts of East Lothian, snorted the *Scotsman* in 1841, 'Hundreds of grown-up colliers . . . never enter a place of worship, but spend the sabbath in the vilest debauchery and rioting.'[115] Yet it does seem that chapel-going was gaining at the expense of church-going. And it seems that even in the 'worst' colliery districts the number of regular attenders, at church, chapel and citadel, was not much lower than in the rest of the country and was far higher than is common anywhere today. The only exact evidence comes from the so-called religious census which showed that between 20 and 30 per cent of miners and their families worshipped on Census Sunday, 30 March 1851. In the colliery districts of Durham for example over a quarter of the population went to a religious service of some sort, with more than a sixth worshipping at their local chapel.[116] Other evidence points in the same direction. An 1856 survey of 1,600 Butterley Company employees living in East Derbyshire confirms that between about 15 and 20 per cent of those mining families were chapel-goers.[117] Attendance may have been lower in the towns and cities but in few coalfields was there any sign of that general weakening of nonconformist influence which can be detected in other parts of the country after the turn of the century.[118] Yet it is easy to be misled. Most miners did not wish to worship;

and not all those who wished to were able to. Many surface workers had to work on Sunday and some men, like Ralph Eckersley of Wigan, refused to take their families to church because they were unable to afford decent clothes for the children.[119] It is important to appreciate that despite the generalisations commonly made about miners and nonconformity, the great majority of the colliery population did not go to chapel.

If it is hard to estimate the precise extent of Methodism in the coalfields, it is even more difficult to determine the effect which the sect had upon behaviour. This is a problem which has been the subject of intense historical debate.[120] It is all too easy to exaggerate the extent of Methodist influence. Methodist sympathisers have rather tended to misrepresent the conditions of the unconverted in order to emphasise the ameliorative impact of their faith. At the same time even the most hostile of historians may be tempted to overestimate the effect of the new religion because its growth coincided with a period of intense, though not necessarily related, social change.[121] Some miners who went to chapel showed no particular hostility towards the Established Church and no special affection for nonconformity. They went to church or chapel as convenient but felt all the time that the Established Church was somehow superior. 'I shall take all my children to the Church and have them baptized,' remarked a Derbyshire miner's wife in 1856: 'They have been baptized by the Ranters but I don't think it's right.'[122] A Fife miner left the Church of Scotland in order to join the Plymouth Brethren who met in Cowdenbeath. He returned after a few months explaining that 'after a while he got tired of religion and he had to come back to the Auld Kirk, as the Church of Scotland was best known by him.'[123]

What effect did Methodism have upon the miners' behaviour, and especially upon their drinking habits? For some, no doubt, 'the new religious life took the place of drunken carousels.'[124] A miner with nine children at the Brandon colliery in Durham had always been a wasteling. Suddenly he was converted and the transformation amazed his friends and acquaintances.

> I was then shown into their front room, and it quite dazzled my eyes to look upon such a change. There was a neatly furnished room with a light carpet and centre table, on which was placed several books. The walls were adorned with kinds of pictures . . .[125]

For many, however, Methodism could only moderate, not eliminate, the attractions of the beerhouse. The reformed drinker was always likely to be tempted by his former companions when pay-day came round.[126] Indeed it has been suggested that the Methodists' adoption of total abstinence from about the 1830s may have weakened their influence by widening the gulf between themselves and the mass of the miners.[127] Nonetheless it does appear that Methodists became disproportionately influential in modifying behaviour, at least in the 'typical' isolated mining communities to be found in parts of South Wales, Durham and Northumberland. Even if Methodist teaching itself was unfavourable towards independent working-class activity, Methodist organisational structures were not. With its classes, meetings, discussions, committees and fund-raising activities, the chapel became a school for self-help. The weekly class meeting, asserts Jim Bullock, 'became a superb training ground for public life, for these same men might develop into local preachers, and there they would *really* find their voice and confidence. These men were a splendid example to their fellow men, expressing themselves freely and becoming influential figures at work.'[128] According to a hostile Derbyshire critic, Methodist

> leaders are to be found in Co-operative Societies, or Hospital Committees, among Chapel Trustees, and Deacons, in Teetotal Societies, members of Parish Councils, School Boards, and a variety of other places. No-one can do anything but themselves. If they could rule, earth would be a paradise, and, because they cannot, injustice and misery are the consequences . . . These people you meet wherever you go, but in these colliery districts they stand out more prominently than anywhere else as they are men who have studied well the science of knowing what to say to catch the average collier's ear. The moment the collier begins to think for himself their power is gone for ever.[129]

Whether it be seen as benign or as malignant, Methodist involvement in working-class organisations was an undisputed fact of late nineteenth-century coalfield life.

Methodism exerted an influence which extended far beyond the confines of the congregation. Some non-attenders took part in chapel social life while a good number were married or related to members. Towards the end of the period about a sixth of those living in the Deerness Valley in central Durham belonged to either the Primitive or the Wesleyan Methodists; yet about 40 per cent of all households were commonly regarded as being Methodist.[130] For believers and non-believers alike, churches were a major social institution in the nineteenth century.

In the mining village of Trebanog, as in most mining communities, the people could be separated into two main streams — those who went to chapel and those who did not. Most of my friends during this period [just after World War I] were 'chapel boys', not because of any deep religious conviction but out of deference to our parents, in my case to mother, who was deeply religious. I went to chapel three times on Sunday, and Band of Hope prayer meetings and Young People's Guild on three nights a week. It was accepted as a duty and on more than one occasion my brother and I were called from cricket or football to attend such week-night religious occasions.[131]

Here then was one way in which the chapel's influence became all-pervasive. Another, hinted at in the passage above, was by providing families in many towns and villages with their only real alternative to the pub, a new working-class culture based upon the chapel.[132] At places like New Lambton in County Durham, the 'little unassuming Methodist chapel was the only place of worship, and its Sunday school the only place of education, in the village for more than sixty years of its history.'[133]

The chapel anniversary and Sunday school treat were big events in the lives of the community.[134] Yet it was not the special occasion but the regular, day-to-day function which cemented the faithful together and provided the miners and their wives with a new focus for their social and cultural life.

There was something for almost everyone: penny readings, love feasts, temperance classes, Pleasant Sunday Afternoons, choir practices, 'tonic sola-fa' classes, Sunday school classes, concerts, Band of Hope sessions and a whole range of fund-raising activities. You could spend every night of the week at the chapel and most of the weekend too. The Bowers Allerton Mission Hall near Castleford

> played a very active part, not only on Sundays and festivals, but every night of the week.
> On Monday there was the Christian Endeavour which was chaired by the undermanager, Mr Robinson. They sang hymns and had addresses, and once every month one of the members was encouraged to speak. This led to discussion amongst the members. On Tuesday night there was a Chapel class meeting. This was a real highlight of the week, a spiritual feast, since all the men and women who had given their hearts to God were eager to give a public testimony to the difference God had made to their lives . . .
> Wednesday night was Mother's night at the chapel, and a baby-sitting night for the men. What the women talked about at these meetings, we never knew, but what we do know is that arrangements were made for chapel teas and other social functions, as well as for visits to the sick and aged. I do know that my mother always came home looking relaxed, happy and more contented than when she went.
> Thursday night was our Band of Hope Meeting, and again this was founded and invariably chaired by Mr Robinson, the undermanager. This was a teetotal affair, and all young men and women had to sign the pledge never to smoke and never to drink intoxicating liquor . . .
> Friday night was the big night for men, because this was the Men's Friday Night Bible Class . . . You became a member of this class when you left school and could remain a member until you died . . .
> Saturday night was a totally different night at the chapel, because on Saturday night we had the chapel socials and other innocent get-togethers. Even on these pleasurable occasions we always opened with a hymn and a prayer.[135]

For women in particular this was a most important development. Whereas working-class men had always been able to enjoy a range of activities based upon the pub, their wives had been forced to create what leisure they could within the narrow confines of family life, by gossiping, popping next door or running round to the corner shop. For women the distinction between work and play was a tenuous one. Now, just as families were getting smaller and real income rising, there began to emerge a new, non-drinking, non-gambling focus for social life. Religion and studying were never so popular as drinking and fighting and it would be absurd to pretend otherwise. Yet all too often attention has been focused on the hard-drinking, hard-fisted miner to the complete exclusion of his opposite number, the serious, ambitious men hurrying from the pit to the reading room or the public lecture hall. Neither was typical but neither should be forgotten.

7

The Miner and his Clubs

Not least among the failures for which the miners have been castigated over the years has been their 'inability to save', their failure to prepare financially for the future.[1] All observers are agreed that the nineteenth-century miner was improvident. 'To economy,' complained a Scottish minister in 1798, 'he is, in general, an utter stranger.'[2] He just would not save even in good years. As the *Derbyshire Times* commented in 1861: 'It is to be regretted that the mining population generally do not, in these times of prosperity, carefully husband a portion of their earnings, to be laid by to fall back on in adverse times.'[3] The boom of the 1870s brought forth a barrage of complaints. The *Colliery Guardian* regretted that 'On account of the great prosperity that prevails throughout the mining districts at present, the habits of a great many of the working classes are becoming terribly loose and demoralised.'[4] Or again, 'Provident habits being very rare among the mining population, they are spending their present abundance in useless indulgence and intemperance.'[5] Indeed it was widely believed, even in the colliery community itself, that few families were able to put by for a rainy day. 'It is all very well to talk of saving money,' pointed out one 'Poor Married Man', 'but it is not so easy to do.'[6]

Modern social and labour historians scarcely mention either miners' savings or miners' insurance. But it is assumed that the twin fears of physical and financial insecurity so discouraged thrift and foresight that in any domestic crisis — be it accident, sickness, old age, unemployment, short-time, a strike or a lock-out — the miner was thrown onto the tender mercies of his employer, the parish or of the charitable public.[7] Even the best historians of the coal industry fall

172

into this trap. Dr J. E. Williams thought that 'There was, no doubt, a streak of Epicureanism in the miners' philosophy,' and that 'The question of the miner's inability to save is linked with his propensity to "play".'[8] Dr Challinor believes that the Poor Law served as the major source of compensation for mining injuries: 'the average miner could reasonably expect to be involved in several accidents during the course of his working life. If unfortunate, he might be crippled or killed. As a result, his family would become paupers.'[9] The reluctance to save is said to persist even in the post-nationalisation era. Despite

> the continual danger of violent death [and] a constant risk of injury . . . the miner feels . . . that all the savings he could possibly muster would make little financial difference if he was seriously incapacitated and rendered permanently incapable of work. Spread over a period of several years the spending of the savings would scarcely help . . . The tendency, therefore, is to give up saving as a bad job, and live from day to day, spending of the money as it is earned in the belief that 'they'll manage somehow' come what may.[10]

Never is it suggested that the nineteenth-century miner's own foresight might enable him to support himself and his family without recourse to charity or to the parish.

Certainly the Poor Law authorities did deal with a vast amount of distress. The Overseers and later the Guardians helped the sick, the injured, the inadequate, the aged, the widowed, the orphaned and the abandoned.[11] In times of particularly severe depression, miners and other workers in England and Wales were set to work harvesting, road-making, stone-breaking and oakum-picking.[12] In the winter of 1879-80, for instance, unemployed Dean Forest miners were sent to do stone-breaking under the direction of the Monmouth Board of Guardians.[13] On occasion some Guardians were even prepared to relieve the hardship caused by strikes. During the 1898 lock-out in South Wales, the Bedwellty Guardians opened relief works and labour yards and even paid some of the men's debts.[14]

Not all applicants for assistance were so fortunate. The

Scottish Poor Law invariably refused any relief to the able-bodied so when for example the 1842 Act displaced seventy-three women from the Bannockburn colliery in Stirlingshire, 'Some few families were reduced to a state of destitution, which was very partially relieved by the parish allowance, the able-bodied not being permitted to share in it.'[15] In some parts of the country, local bye-laws forbade the payment of relief to anybody of 'immoral habits'.[16] A more common restriction was the refusal to assist widows with only a few children. The invariable rule in the Barnsley Poor Law Union was that, although the Guardians always supported small children, widows were expected to work. Nobody, it was confidently asserted, thought that assistance should be offered to a widow who was able to earn her own living.[17] Very often, too, outdoor relief was refused when it was believed that the dependants of the deceased had failed to use their friendly society funeral grant in a responsible manner. In 1880 the Wolverhampton Guardians turned down a widow's application for outdoor relief when it was learnt that she had spent over half the £16 received from an insurance club after her husband's death in order to pay for his funeral. Instead the widow and her two children were offered the workhouse.[18]

This was relatively unusual for it is a misconception to suppose that the hated workhouse was the usual way of dealing with paupers. Most applicants for financial and medical aid were offered outdoor relief. But, 'There was nothing particularly humane about the outdoor relief system . . . Overseers and Guardians used it as a means of getting rid of applicants for relief at the least possible cost in time, trouble and expense to themselves and the ratepayers.'[19] The medical treatment provided was rarely sufficient to effect a cure and the cash allowances paid were derisory. In 1871, for example, the Wigan Board of Guardians was paying the widow and three young children of a sober and industrious collier a mere 3s. (15 pence) a week with the result that she was compelled to sell her wedding ring and other personal belongings. She would have been little better off anywhere else for payments never rose above 3s. a week for adults and 1s. 6d. (7 pence) a week for children. So the family of even an oncost worker in one of the poorest coalfields would notice a sharp drop in income if they were forced to go on the parish.[20]

Most Guardians sought to encourage the working population to become independent of the Poor Law. To refuse relief to the thrifty miner on the grounds that he was receiving assistance elsewhere was unlikely to persuade him to make provision for himself and his family. Accordingly many Boards ignored central Poor Law policy and took account of only part of any allowance which the miner or his dependants enjoyed from a friendly society or similar institution.[21] Other Boards showed their approval of the thrifty miner by granting loans and by the manipulation of indoor and outdoor relief. Restrictions were sometimes placed on the offer of outdoor relief if it was felt that the applicant had made insufficient preparation for the future. When in 1881 the Salford Guardians were confronted by two miners suffering from slight accidents, it was decided to grant them out relief but only on condition that the money was repaid when they resumed work. 'The colliers ought to know', remarked the Relieving Officer, 'that the workhouse was not a sick club for them.'[22] In 1887 in the Forest of Dean, the Guardians of the Westbury-on-Severn Union decided to offer the workhouse to an injured miner since, although he had no children, he had made no provision at all for his old age.[23] Two years later, the Barnsley Board offered only indoor relief to the wife and five children of a collier injured at the Carlton Main colliery the day after leaving a local friendly society, the West Riding Miners' Permanent Relief Fund. This policy was adopted, the widow was told, because the Guardians 'considered her husband had acted diametrically in opposition to the interest of the ratepayers by throwing up the chance of relief.'[24]

It is clear then that in every coalfield the Poor Law acted as a leading agency of relief. But it was not the only one. Charity too played an important part. Soup kitchens and relief schemes of various sorts were opened whenever distress became widespread.[25] In the severe depression of 1878, Sir Horace St Paul of Ellowes Hall, Dudley, agreed to have about thirty acres of colliery waste lands at Coseley cleared and cultivated in order to provide work for the local unemployed.[26] In the following year, 460 Brynmawr children 'assembled in the National School, and . . . received a wholesome and satisfying breakfast of coffee and bread.'[27] Help was also forthcoming during strikes and lock-outs. During a dispute at Oldham in 1858, local tradesmen and others entertained 612 miners and their families to a dinner of boiled potatoes, a hundred plum puddings (weighing 5 lbs [2.2 kg] each) and a roasted ox which cost £21 to buy.[28] It was normally relatively easy to obtain credit from local shopkeepers, some of whom would be related to the strikers and many of whom would depend on the miners' custom once the dispute was over. Where ties of blood and self-interest did not suffice, the threat of violence might. A Standish grocer who refused to give credit during the Lancashire strike of 1874 soon found that his hay-rick was burnt down.[29] Another expedient, popular in the early nineteenth century, was to go from town to town 'dragging a waggon load of coal, and soliciting relief'.[30] Yet another was to evade the charge of begging 'by offering hymns and rhymical appeals for aid . . .'[31]

It was after mining accidents however that the public responded most readily to appeals for assistance. Occasionally subscriptions were raised for the victims of day-to-day accidents. In 1861 twenty pounds was collected at a concert held in the Central Hall, Blyth for a Northumberland miner who had been seriously injured at the Cowpen colliery and in 1895 a committee was formed to raise money for a man who had been blinded while working at the Brynn Hall colliery outside Wigan.[32] Although the high death and injury toll among miners was overwhelmingly the result of such incidents as these, it was only the relatively few major disasters which really attracted the attention of the charitable public.[33] On these occasions huge sums were sometimes raised. The Hartley fund of 1862 raised £81,000, the Blantyre fund of 1877 nearly £48,000, the Clifton Hall fund of 1885 £27,000, the Thornhill fund of 1893 £36,000 and the Senghennydd fund of 1913 over £126,000.[34] Yet these colliery disaster funds shared many of the weaknesses common to all forms of nineteenth-century charitable activity. They were not as successful in directing assistance to those most in need as a mere catalogue of the sums raised might suggest. They helped the families of no more than an eighth of all miners killed in the industry. Often too the subscriptions raised bore little relation to the amount of distress to be relieved and the benefits provided were far too low to prevent destitution. Even when vast sums of money had been raised, payments remained low. Twenty years after the Oaks explosions of 1866, a surplus of over £29,000 was left to provide thirty-one dependants with 'weekly sums of six shillings and less'. Indeed within two years of the extablishment of the Hartley Colliery Relief Fund, the executive committee realised that nearly a quarter of the money subscribed would not be needed to pay the widows 7s. (35 pence) and their children from 2s. 5d. (12 pence) to 3s. 6d. (17 pence) a week. But the allowances were not increased. Even payments of this magnitude were commonly felt to be too high. At a meeting of the Caerphilly District Council a few weeks after the explosion at Senghennydd, 'Mr Mark Harding said he had seen the grocery bill of a person who had received a grant of 10s., and this bill included a quarter of a pound of tobacco, two pounds of bacon at 1s. 3d., butter at 1s. 3d., and jam at 10d. a pot. He did not think these were necessary things.'[35]

The employers too helped miners and their families who were in need. Some treated the old folk, some were prepared to advance loans, and a few even helped their employees while they were on strike.[36] But it was after mining accidents that the employers, like the public generally, chose to make their greatest contribution. The Durham and Northumberland owners and in other coalfields certain individuals, like Lord St Oswald in Yorkshire and the Earl of Dudley in the Black Country, always made systematic provision for those injured at their pits.[37] A few concerns like the Buccleuchs, Dundases and Clarks in the Lothians, the Countess Waldegrave in Somerset and the Lambtons, the Londonderries and later the Pelton Colliery Company in Durham, granted permanent pensions to widows and former employees.[38] A tiny handful went on to provide what anounted to a complete system of social welfare. The miner employed in South Yorkshire by the Fitzwilliam family enjoyed many benefits:

if injured his wages were continued, or more usually an

allowance made, and free medical attention was provided. When the state of trade did not justify employment in his usual occupation he was often found other work to do . . . if he died his family were allowed to live on in their home and his widow was granted a pension. On retirement he was added to the list of weekly pensioners . . .[39]

These were the exceptions which proved the rule. The vast majority of employers did nothing more than provide those injured or 'worked-out' with a light job and some medical assistance, bury those who were killed, and subscribe towards the disaster fund which would probably be opened if they were unlucky enough to have a major accident at one of their collieries.[40] Like the help forthcoming from members of the general public, the assistance provided by the owners was always greatest after major disasters. Nor did it necessarily reach those most in need for there is no doubt that the owners manipulated their charitable activity as yet another weapon in their already powerful disciplinary armoury. In Northumberland and Durham it was the custom to allow the 'relicts' of men killed in the mine to retain occupation of their colliery houses. Yet, just sixteen months after the Seaham explosion of 1886, thirty-seven widows had given up their homes at the request of the management and eighteen months after the Usworth explosion of 1885, Bowes and Partners gave the widows four weeks notice because their homes were required for other workmen.[41] The medical attention provided by the employers also came in for fierce criticism. In October 1859 a boy was burnt at the Robert Town colliery near Leeds and 'Although the burning happened on Friday, the proprietors of the pit (Messrs Parkins) sent no medical relief to the unfortunate sufferer until the following Monday; and death ensued not long afterwards.'[42] Eight years later it was claimed in Yorkshire that delay on the part of the Edmunds' Main colliery doctor had resulted in the death of an eighteen year old youth.[43] In fact as late as 1891 the Northumberland Miners' Mutual Confident Association was complaining of delays in the provision of treatment to their members.[44] There were other causes of discontent in the North-East of England where the miners' accident allowance (or smart money) was easily manipulated. At many pits the owners were entitled to refuse payment to an injured man when he left their employment. And many owners claimed that a strike constituted a termination of contract. In 1831, for instance, seventeen year old Thomas Lawton was burnt by an explosion in a South Durham pit and laid up for six months. For some weeks he received his 2s. 6d. (13 pence) smart money but then his colleagues came out on strike and his money was stopped.[45] Other employers were far less subtle: they simply sacked the injured. In 1853, as part of their struggle against trade unionism in Northumberland,

> The colliery agents at Seaton Delaval discharged upwards of thirty men . . . Mr. Edward Richardson . . . had a son Matthew who was hewing. He was brought home almost lifeless, so much injured that there were men appointed to attend him night and day for some months. When this unhappy event occurred, he had gained strength to go out on crutches. He lived in the house with his father, and as Edward Richardson was one of the thirty who were discharged, it resulted in the son's smart money being stopped.[46]

It was not only by their own direct assistance that employers helped miners and their families to withstand the difficulties brought about by accidents and other misfortunes. In every part of the country strenuous efforts were made to compel — or when this became impossible, to encourage — miners to insure themselves against death, disablement and sickness. During the first three-quarters of the nineteenth century, owners everywhere organised colliery insurance funds, membership of which constituted a condition of employment. These funds usually insured their members against accident, often insured against the costs of medical attention and sometimes insured against ordinary sickness.[47] Some insured all three. In the 1840s every employee of the Butterley Company in Derbyshire whose wages amounted to 8s. (40 pence) a week, was required to subscribe a shilling (5 pence) a month to the company club. In return he received 6s. (30 pence) a week when ill or injured as well as free medical attendance in all cases of serious accident.[48] Some funds provided even wider benefits. The men and boys em-

ployed by W. P. Morewood at nearby Swannick and Summer-
cotes were allowed 6s. (30 pence) a week for accident, 3s.
(15 pence) for illness and 2s. (10 pence) when past work —
and all for the payment of a shilling (5 pence) a month.[49] In
Clackmannanshire the Alloa colliery fund was managed by the
colliery accountant; the men paid 8d. (3 pence) a week and
the benefits included school fees, medical care, christening
bonuses, and allowances for widows, the sick and those unable to work because of old age.[50]

For three-quarters of the nineteenth century, these com-
pulsory pit clubs afforded miners their major sources of
compensation after non-fatal accidents. But they were open
to serious objections. Payments were rarely more than 8s.
(40 pence) a week, less than most hewers were able to earn in
a single shift at the end of the period. And these payments,
low as they were, were reduced even further when the re-
cipient had been on the funds for a year or so.[51] Nor was this
all. Although it was the men who paid into the funds, they
had little control over their management and were rarely
allowed to inspect the accounts.[52] Not surprisingly, and
quite correctly, the men suspected that many employers were
milking the funds.[53] It was calculated that the men's subscriptions to the Clay Cross field club in Derbyshire during the
1860s totalled over £5,000 a year, out of which the doctor
received only £400.[54] Meanwhile at Messrs Charlesworth's
pits in West Yorkshire the men claimed that 'the rates of contribution . . . leaves a good profit to the employer, who
pockets the same and gives no account.'[55]

As if all this was not enough, many funds were actuarially
unsound. A good number divided up any remaining surplus
at the end of each year.[56] By their very nature, these funds
were organised at colliery level so that the solvency of each
fund depended on the solvency of the individual employer.
Closure or a change of ownership left the injured and bereaved
without financial support except on those rare occasions
when the men's contributions had been capitalised and invested elsewhere. A delegate to a union conference at Tipton
in 1870 told how his brother had been badly hurt the very
week that a pit closed; so although he was ill for six months
he never received a penny from the pit club to which he had

been a subscriber.[57] The narrow scale on which the funds operated also meant that they were unable to cope with a large accident. Thus the Oaks colliery fund collapsed from the great pressure placed on it by the explosions of 1866 and although the Clifton Hall colliery fund in Lancashire enjoyed the reputation of being one of the richest pit clubs in existence, it proved quite incapable of meeting its liabilities when 178 members were killed in 1885.[58]

There were other drawbacks. Like the friendly societies promoted by the railway companies, most funds refused relief to men who left their jobs for any reason, no matter how long they had been subscribing.[59] Many funds failed to provide even the relief that was promised. A lot of 'men find, when they are permanently injured, that their expectations of support are cut short after a very few weeks' receipt of relief.'[60] Some employers manipulated the pit clubs, as they did their direct charitable activity, as an instrument of industrial discipline. 'Given, a local relief fund, managed by masters . . .' explained 'A WORKMAN', 'the miner's widow and children might suffer for a presumed insubordination on the part of the poor fellow that was gone.'[61]

Where pit clubs provided medical treatment, the doctors, who were appointed by the owners but paid by the men, invariably came in for a great deal of criticism. The men always believed that club doctors were over-eager to get them back to work. It was said at Coxhoe, County Durham that if a man was in bed with a broken leg for five or six months and then got up on crutches, the doctor would immediately tell him to return to work.[62] Some doctors lived miles from their patients, some were unqualified and some just plain incompetent.[63] A few examples will suffice. One of the Children's Employment Commissioners found in 1842 that with a single exception, not one colliery doctor in Northumberland or Durham kept any records at all.[64] It was not unknown for doctors to refuse to treat an injured miner. In 1863 a Shropshire field club doctor declined to treat a broken ankle because he maintained that the accident had been caused by drink. Even when this particular patient obtained a hospital note to confirm that his ankle was crushed, the doctor still refused to take the case.[65] At Coxhoe the colliery doctor refused to come out after 7 p.m. and would send instead his assistant, the unqualified son of an underviewer.[66] Sometimes when treatment *was* provided, it did more harm than good. The *British Miner* claimed in 1863 that at Consett 'Holloway's pills and ointment reign triumphant'.[67] The following year, five men were injured near Durham and although the colliery doctor attended one of them for a fortnight, he concluded incorrectly that there was nothing wrong with a damaged shoulder and set another man's broken arm so badly that it had to be rebroken.[68] At the turn of the century the young Tom Williams was kicked in the face by a pony at the Warren Vale colliery in South Yorkshire.

> The colliery trap rattled me off to the doctor, who had a quick look at the damage and gathered up his needle and wire or whatever it was. I sat in a chair and shivered and waited for the stitches. The engine-wright and a blacksmith, both sixteen-stoners, stood behind me to see that my head and arms stayed still while the doctor put the stitches in. There was, of course, no question of a local anaesthetic being used.[69]

The employers, the general public and the Poor Law authorities all helped miners and their families to survive periods of accident, sickness, bereavement, old age and lack of work. But even at its best this help was likely to prove completely insufficient. The greatest amount of assistance was always forthcoming after major disasters. After the Oaks explosions of 1866, the pit club collapsed. But the owners allowed the widows to live rent free in colliery houses, the disaster fund paid them 5s. (25 pence) a week plus 1s. 6d. (7 pence) per child, and the Barnsley Guardians added 2s. (10 pence) a week for each child.[70] So after the greatest mine disaster before 1913 the most that a widow with two young children could possibly receive from the combined efforts of the public, the employers and the Poor Law was free accommodation and 11s. (55 pence) a week. Seen against this background it is not difficult to understand why the nineteenth-century miner came to belie his reputation for improvidence and lack of thrift. Miners' savings took many forms. An earlier chapter showed how some families bought their own homes. Others used the co-op, keeping 'a few pounds in the store', as a first line of defence against illness and unemployment.[71] Many more belonged to the slate clubs, Christmas clubs, holiday clubs, clothing clubs and money clubs which were organised in colliery districts as everywhere else. Money clubs for example were organised in alehouses: weekly subscriptions were collected over a predetermined period and members either received their savings at the end of this time or lots were drawn and each week one member would receive the whole of that week's subscriptions.[72]

As the century wore on, the ability to save became a feature of working-class life and the miners turned increasingly to more formal ways of saving.[73] Occasionally a miner invested in his local bank or building society. In 1863-4 a waggonman and two banksmen between them had £40 6s. 8d. (£40.33) deposited with the City and County of Durham Permanent Benefit Building Society and a Clay Cross miner, William Elliott, lost all his savings when the Chesterfield Bank collapsed in 1880.[74] More popular were the savings and penny banks, which, as their names suggest, were prepared to accept very small sums for deposit. Trustee savings banks were opened in many coalfield towns: in Wigan in 1821, Sunderland (1824), Blackburn (1831), Dunfermline (1837), Kirkaldy (1838), Bilston (1839) and Falkirk (1845).[75] Within fourteen months of the opening of the Dunfermline bank, 270 'Coal hewers, miners, quarrymen and labourers' had opened accounts.[76] Miners in the East of Scotland continued to save. A 'considerable minority', reported the *Colliery Guardian* in the prosperous early 1870s, 'are really bettering themselves and their houses, and taking advantage of savings banks besides.'[77] Penny banks were also well patronised in the coalfields. The Yorkshire Penny Bank was opened in 1859 and by 1886 it had branches in such colliery villages as South Kirkby, Ryhill and Hemsworth, the latter having 1,132 accounts totalling £1,182.[78] The habit of saving was becoming more deeply entrenched than is generally realised. So when the young Lancashire boy Richard Ridyard started work for the Abram Coal Company in 1874, his mother made him deposit 2s. 6d. (13 pence) of his first fortnight's pay with his Sunday school savings bank.[79] Such foresight was unusual but, by this time, not exceptional.

Given the dominance of men in the life of nineteenth-century mining communities, it is not surprising that few families put by for the wife except to meet the costs of confinement. Thus it was said in South Staffordshire in the 1840s that while colliers' wives saved, sixpence by sixpence, the 10s. 6d. (53 pence) needed for their confinement, their husbands would often find the money and spend it so that the doctor never in fact got paid.[80] However, some women joined their own friendly societies and some men, like those at Merthyr, did insure their wives specifically against the cost of midwifery.[81] As might be expected, it was the man, the breadwinner, that mining families took the greatest trouble to protect. They insured him against sickness. The simplest way was for the men themselves to organise a whip-round; at Darfield Main in South Yorkshire in 1893 the workmen organised a smallpox levy of sixpence (3 pence) a head so that any contributor taken ill with the disease would be paid 25s. (£1.25).[82] There were also 'benefit societies which contract for so much a head with medical men' to secure treatment as soon as it became necessary.[83] And there were funds offering more substantial sickness benefits. At Wednesbury in the Black Country the Rev. T. Clarkson began a sickness fund which paid out a death grant as well as 10s. (50 pence) a week during the first six months of sickness, 7s. 6d. (37 pence) a week during the second six months and 5s. (25 pence) a week thereafter.[84]

Sickness insurance was important but it was by insuring against the industry's day-to-day — indeed hour-to-hour — accidents that miners made their greatest effort to remain independent. The miner's life was physically dangerous and financially insecure. Mining communities, aware of the precarious margin between survival and ruin, were becoming more thrifty and self-reliant. The general idea of the miner as a spendthrift then becomes a travesty. Through their friendly societies and trade unions, miners supported those local voluntary hospitals that dealt with accident cases, sometimes trying to adapt them specifically to meet their own needs. In 1880, the men at Lumphinnan's colliery in the west of Fife began to raise the £200 that was needed to open a miners' surgical ward at the Edinburgh Royal Infirmary.[85] The problem of inadequate medical attention was tackled in other ways as well. Doctors were hired to remain at collieries throughout the whole of the working day and specialised clubs were formed, like the Monckton Main Convalescent Fund in Yorkshire and the Golborne and Lawton Medical and Surgical Aid and District Nursing Association across the Pennines in Lancashire.[86]

In every coalfield in the country the usual way of insuring against accidents was to join one or more of the friendly societies which promised subscribers financial relief and (perhaps) some medical attention.[87] Throughout the whole of the nineteenth century most miners belonged to at least one local accident insurance club, and many belonged to several. The oldest, simplest, and for many years the commonest, type of club was that which organised a collection whenever a fatality or serious injury occurred. Typical was the procedure at Williamson's Soft Pit at Babbington in Derbyshire: when the young Joseph Aram was injured in the 1840s the men and butties collected among themselves for over nine months in order to make him a weekly allowance of 3s.6d. (17 pence).[88] Fifty years later a 'gathering' of a shilling (5 pence) a head was still being made at the Bestwood colliery in Nottinghamshire so that permanently injured workmates could try new careers as hawkers.[89] 'The colliers', remarked the clerk to the Walsall Union, 'are exceedingly kind, and make collections for any of their number who may be injured.'[90]

When we had counted the hard earned shillings and made sure that we had got them all we filed out at the opposite end [of the pay office] from which we had entered. A man with a freshly bandaged stump of an arm, smelling strongly of hospital odours, stood at the door with doffed hat begging. His body that once was strong and able to do its share of work was wasted and weak and his face bore the mark of suffering. Many a penny dropped into his hat that day athough the men don't coutenance such bare-faced begging as a rule. He had been a collier. Not one of us could tell whose limb might be missing ere the next pay day.[91]

Also popular were societies which levied their members whenever funds fell to a predetermined level. Thus in Yorkshire the treasurer of the Old Silkstone Collieries' Accident Society was required to stop a contribution of 6d. (3 pence) a month (boys under sixteen 3d. [one penny]) when total reserves dropped to £2.[92] But by far the most popular type of local friendly society was the permanent club with fixed and regular contributions. Membership was high wherever coal was mined. In 1861, for instance, a Miners' Accidental Association was meeting at the Astley Arms Inn at Dukinfield in Cheshire and a quarter of a century later the surface workers employed at Brampton in Nottinghamshire established the Cortonwood Surface Sick and Accident Society.[93] From the 1830s onwards miners all over the country organised and joined such local accident insurance funds. Membership was part and parcel of coalfield life. On pay night, remarked a somewhat surprised observer of the Rhondda scene in 1872, 'many of the miners [were] having to meet, after receiving their pay . . to attend benefit clubs, &c.'[94] That scenes like this were repeated in every coalmining district should come as no great surprise because even in urban slums 'all but the poorest could fall back on that bulwark of the times, the funeral friendly society.'[95]

In challenging the conventional view of the thriftless miner it is unnecessary however to rely just on the evidence of local friendly society membership. The same developments that were improving the quality of family life – declining mobility, fewer children and higher real earnings – tended as well to encourage thrift and thought for the future in other directions. From about the middle of the century there was growing support for the insurance and other facilities offered by more broadly-based friendly societies, by the commercial insurance industry and by the trade union movement.

Commercial insurance grew rapidly in popularity. At mid-century it was a predominantly middle-class activity though by the early 1860s the Friend-in-Need Life, Fire Guarantee and Accidental Death Insurance Company had agents in all colliery districts and numbered among its policy-holders miners from Staffordshire, Yorkshire, South Wales and Northumberland and Durham.[96] Despite growing competition, business was becoming monopolised by large companies like the Refuge, the Pearl, the British Workmen's and, in particular, by the Prudential. By the mid-1880s a quarter, and by the mid-1890s more than half, of all English miners were insuring with this one company alone. By the beginning of this century taking out a policy with the man from the 'Pru' 'had become virtually a universal habit'.[97] A Yorkshire miner remembers that 'We were all insured for a penny a week with the Prudential, so we were sure of getting ten pounds at death, so they could afford to bury us. I think this was an insurance policy everyone in the village had.'[98]

During the second half of the century many miners joined friendly societies that were more broadly-based and hence more stable than the local clubs which were considered above. The East Dean Economic Society often paid funeral allowances to the widows of miners killed in the Forest of Dean. In Lancashire the Blackburn Philanthropic Burial Society paid the funeral expenses of nearly half the sixty-eight men killed in the Altham colliery in 1883 and two years later officials of the Sincere Sick and Burial Society at Manchester found that their 'funds had been drawn upon considerably' by the deaths of 178 miners at the Clifton Hall colliery.[99] The affiliated orders and the great collecting societies both attracted about an eighth of the country's mineworkers. Thus thirty-four of the 361 killed at the Oaks were members of the Royal Liver and during the late 1860s some 25,000 miners subscribed (from 4d. to 8d. [2 pence to 3 pence] a week) to the Manchester Unity of Oddfellows.[100]

But one of the most striking and neglected examples of the thrift of late nineteenth-century coalminers is to be seen in the growing number of subscribers to the miners' permanent relief funds, friendly societies which, while sometimes providing pensions and medical relief, were designed essentially to insure miners and their dependants against the financial loss caused by colliery accidents. The first permanent relief fund opened in Northumberland and Durham in 1862 and by 1881 there were seven covering every major coalfield in England and Wales. Despite early trade union hostility and criticism of both the costs and the methods of administration, the miners came to accept the permanent relief funds as an efficient, if impersonal, system of insurance. The benefits (4s. (20 pence) to 10s. (50 pence) a week after injury, cash payments of between £5 and £23 for death, and weekly allowances of 5s. (25 pence) for widows and 2s. (10 pence) or 2s. 6d. (13 pence) for orphans) were good value for a weekly subscription of between 2d. and 4½d. (1-2 pence). More important, the funds never failed to meet their commitments. Clumsy as the name 'miners' permanent relief fund' might be, it was, claimed the movement's supporters, 'most happily chosen and strikingly appropriate'.[102] More important still to the movement's success was the support of the employers in deducting subscriptions at source.

> Before this system was introduced, lapsed membership was very frequent, owing perhaps in some measure, to the smallness of the payments, members put off from time to time until — from sheer carelessness — they were, by rule, excluded from benefits, and in some instances, widows with large and helpless families were left unprovided for. Where the above mentioned system of deductions prevails, such falling off cannot occur.[103]

In 1880 the movement could claim a quarter of British mineworkers as members; in Lancashire over a half, and in the North-East of England nearly two-thirds of the labour force had joined the local permanent relief fund. By 1890 over 40 per cent of the country's miners were members, with the West Riding society claiming a quarter, the two Midland societies a third, the Monmouthshire and South Wales society nearly a half, the Lancashire and Cheshire fund two-thirds and the Northumberland and Durham society over 90 per cent of the miners working in their respective coalfields.[104] The membership figures alone, however, do not reveal fully the leading compensatory role assumed by the funds in England and Wales. The West Riding of Yorkshire Miners' Permanent Relief Fund was typical. In 1883 the society's membership represented just a quarter of all Yorkshire miners. Even so it relieved the equivalent of a third of those injured during the course of their employment, the dependants of more than 60 per cent of those killed and every one of those bereaved by accidents claiming more than four lives. By the end of the century the West Riding Permanent Relief Fund had relieved seventy-seven aged and infirm members at a cost of £6,154; 438 widows and 883 children at the cost of £62,727; and 68,223 members injured in the pit at the cost of £135,943.[105] Here again there is convincing evidence of the growing thrift of nineteenth-century miners.

By far the best known of all the miners' many self-help institutions were of course their trade unions. Indeed, to judge from the proliferation of published and unpublished union histories, it would be easy to imagine that the history of the miners is synonymous with the history of their trade unions. Nothing could be further from the truth. During the first half of the century and beyond, trade unionism was able to make only the most marginal impact on the life of the ordinary miner. The character of the coal industry, the nature of pit work, the heterogeneity of colliery communities, the hostility of the employers and even some of the unions' own policies all militated against the successful organisation of stable association.

Fragmentation and the small scale of most early nineteenth-century production meant that outside Northumberland and Durham there were few sizeable, stable, homogenous groups of workmen to organise.[106] The fluctuations of the trade cycle placed another obstacle in the way of permanent organisation. Union after union flourished during the up-swing, only to flounder during the subsequent down-swing. It is striking in fact that both the best known early nineteenth-century mining unions, Tommy Hepburn's Coal Miners' Friendly Society and the Miners' Association of Great Britain and Ireland, were incapable of surviving, let alone surmounting, a sharp down-swing in the trade cycle.[107]

Nor, as should be clear by now, were mining districts the homogeneous communities which it is sometimes assumed. There was a sharp division, for instance, between the hewer and the oncost worker, a wide gap between the young man at the height of his earning power and the old man seeing out his days on the surface. And union policies did much to aggravate this division. For most of the century the unions were dominated by the hewers. Hardly any effort was made to unionise oncost men; often indeed they were explicitly excluded from membership.[108] So, like mechanics at Normanby in Northumberland, they refused to pay their subscriptions unless 'Bill, the blacksmith, Dick, the sawyer, Harry, the pick-shaft turner, Bill, the tub-boiler, and Joe, the day-man, are made to pay as well.'[109] Indeed right up to 1914 and beyond one had to be a hewer even to qualify to stand as a branch secretary of the Northumberland Miners' Mutual Confident Association.[110] Neither is it true that miners always lived and worked together. In Lancashire, where many men were employed in other trades and industries, miners often did not join the lodge where they worked, thus making it hard to pursue a consistent policy and 'almost impossible to find out whether certain men were members of the union or not'.[111] By the end of the century miners in rapidly expanding coalfields like South Wales were often reluctant to keep moving house and chose instead to commute to and from work. Indeed it was said that some South Wales employers preferred to hire men who lived some distance from the pit because their eagerness to catch their bus or train at the end of the shift made them particularly difficult to organise.[112]

Unionism was hindered too by the constant movement of men in and out of the industry. Despite the growing dominance of mining in many coalfields, unemployed colliers continued to look for, and find, alternative jobs. In the late 1870s Nottinghamshire miners 'engaged themselves in the harvest field for less than a common labourer received', while in South Wales, Rhymney miners 'in consequence of the stoppage of pits, are turning their attention to other branches of industry'.[113] Just as serious was the heavy movement of labour into the industry. The English poured into South Wales, where there were also two or three pockets of Spaniards and Italians, and by the end of the period there were about 1,600 Polish miners in Scotland. The Lanarkshire County Union fulminated helplessly against immigrants who accepted work at less than the agreed hewing rates.[114] So severe did the problem become that in 1905 a strike ballot at the Loganlea colliery had to be printed in both English and Polish and two years later it was agreed 'That the new Rules of the County Union be printed in the Lithuanian or Polish language.'[115] But by far the largest group of non-British workers in the mines were the Catholic

Irish in Scotland. They first came as strike-breakers during the 1840s and 1850s and by 1861 there were 93,000 Irish people out of a Lanarkshire population of less than two-thirds of a million. In villages like Quarter and Holytown as many as a fifth of the miners were Irish-born.[116] They continued to exert a baneful, if less direct, influence on trade union aspirations.[117] Their nationality, their religion and their mobility all set them apart from their Scottish workmates.[118] Always active in disturbances, they were less ready to get involved in the discipline and the regular administrative routine which were at the core of permanent trade unionism.[119]

The power and hostility of the employers was also a serious problem. Some of the older, aristocratic owners like the Fitzwilliams in South Yorkshire were able to use a combination of loyalty and financial benefits to dampen the fires of union. During the course of one lock-out, the proudly paternalistic and fiercely anti-union Sixth Earl remarked that he was not prepared to reopen his pits 'until the workmen have learnt that in the long run their own interest and the interest of their employers are one and the same . . . A house divided against itself cannot stand.'[120] Slowly, larger, less paternalistic and more impersonal forms of ownership began to appear. Yet as one barrier to trade unionism started to recede so another began to emerge. The new owners sank large amounts of capital in the industry and were determined to protect their investment.[121] Following the Lancashire strike of 1868 some small owners let the men return at much the same rates as before the dispute whereas big firms like the Wigan Coal and Iron Company were able to hold out for the full 15 per cent reduction.[122] Similarly, that quarter of the Black Country workforce which was allowed to remain at work at existing rates during the 1884 strike was generally employed by small owners in old pits which were unlikely to reopen if they ever closed.[123] A South Yorkshire miner from Denaby Main explains:

You hadn't much trouble at the family pits, Fitzbillies, etc. There weren't much trouble, you know at little pits. You see, there were a stronger set of owners at Denaby, they could rule the roost — I'd heard it quoted that in the bagmuck strike [at Denaby] Buckingham Pope [one of the directors] said that he had a square yard of gold and he'd sink it before the miners would win. He had other collieries, you know, over in West Yorkshire . . .[124]

In all coalfields the employers reinforced their individual strength by joining together to counteract the growing power of the men. The first West Yorkshire Colliery Owners' Association was founded as early as 1830.[125] The Associated Mine Owners of Scotland was formed in 1862 and the Aberdare Steam Collieries Association in 1864, to be followed in 1873 by the much stronger Monmouthshire and South Wales Collieries Association.[126] The Lancashire Coal Trade Association was established in 1877 and the Dean Forest Steam Coal Association in 1889.[127]

At colliery level the employers worked against the trade unions in all sorts of ways. Some were petty, as on the occasion when the Marquis of Londonderry refused to allow members of the Vane Volunteer Artillery Band to take their instruments to a union demonstration on Newcastle town moor.[128] Some were more subtle. Bribes were offered and accepted. According to the Durham union leader, John Wilson,

It cost a large sum to keep the union out of Haswell; but finally the free drink system and the action of the soulless prevailed . . . It mattered not to those who sold themselves. They were favoured above the good men of the colliery. The condition was congenial to the meanness of their nature. Give them beer and let them feel the approval of the official, and they had no longing beyond.[129]

Agitators found they were offered jobs as officials or, like Arthur Horner, put to work where they had little contact with the other men and put on afternoon shift so that they were unable to attend union meetings.[130]

Most owners were far less subtle. Throughout the century they used the fear of victimisation to keep unionists and potential unionists quiet and in order. A Yorkshire miner remembers that even at the beginning of this century

To contradict the boss, publicly [at the Men's Friday Night Bible Class] was looked upon as something of real significance and would become the talk of the village for the next few days afterwards. But it was not unknown for the boss to take his own action, for on Monday morning he would visit the contradicting miner in his working place and would enquire if he felt as clever that morning as he had done on the Friday evening; if so, he could probably tell the manager why his timber was set out of distance. The manager would then take his tape out of his pocket and measure the distance between the props. Invariably the manager was right, the props were two or three inches too wide apart. The manager would then look at the miner and say, 'Perhaps you will be clever enough to tell me why I shouldn't fine you ten shillings.' There was never any appeal.[131]

Or again in South Wales,

Some men went without what was due to them because they feared victimisation. Another man who was afraid to ask for what were his dues kept count of the amount he was short, and seemed to get some satisfaction by telling us, confidentially, that the company already owed him about forty pounds . . . There are so many ways to rid themselves of an awkward man. It is a long distance in to some of the workings, and stones or slags are often piled on the side of the roadway. Just a few stones pushed into a tram of coal in the darkness means a case of filling "dirty" coal when the tram comes to daylight. Dirty coal means lost orders. How can a committee defend that case? How can the man prove he did not fill it? Or a collier may be kept so short of rails that he has to throw his coal four or five yards farther than the man in the next place. He may be kept short of trams or timber, or put in a place were the coal is very stiff or there is a lot of water. The man does not – he cannot – fill so much coal as those in the other places, therefore he is lazy, is not trying; and with his job goes his house very often. The tied-cottage system of the farms has its ugly sister – the house owned by the colliery company.[132]

Certainly trouble makers were soon weeded out. The eight or

nine union men who began to recruit members at Whitehaven in 1831 were immediately sacked.[133] Even serving on a deputation had its dangers. In 1892 a deputation of twelve men waited on the management of the Carlton Main colliery in South Yorkshire; within a month, eleven of the twelve had been stopped.[134]

That 'ugly sister' of the 'tied-cottage system of the farms ... the house owned by the colliery company' was specially common in Scotland and in the North-East of England. Living in a colliery house brought with it the threat of eviction, a threat which many owners were not slow to use in the struggle against trade unionism. In 1863 one Ayrshire man had his door and windows removed while at Lugar a cartload of the 'foulest manure' was shovelled into the home of a widow who refused to leave voluntarily.[135] Sixteen years later every member of the Durham Miners' Association working for Messrs Heley at Craghead was evicted from his home and it was decided to employ no more union men in future.[136] Miners working for the Wishaw Coal Company of Glasgow were evicted during the course of an 1880 strike against a threatened reduction in wages.[137] During the Denaby bagmuck strike of 1902-3:

> The miners who were tenants of colliery houses had been turned out of their houses and the company boarded up the windows. Families had to find anywhere to live: in the churches, chapels, with relatives or friends who resided outside colliery houses, in tents, in the open fields. It was a terrifying time for our people in the fields or anywhere they could go; then they [the company] introduced blacklegs, and brought in the Metropolitan Police to look after the blacklegs. This did not stop the women and children from throwing stones or anything they could find to pelt the blacklegs. The miners' families howled at the blacklegs and they were militant, telling them to leave the pit and go home. There was a great number of tussles with the men as they proceeded to the houses which the company had provided. These were all in one block, to keep them under the protection of the police.[138]

These were the sorts of lengths that a determined employer would be prepared to go to in his fight against the local union. With such hostile employers and a fragmented labour force, it is not surprising that stable trade unionism took so long to find a firm footing in the coalfields. Even at the start of the twentieth century many men, like those at Seven Sisters in the Rhondda, were too scared to pay their union contributions at the colliery office, preferring instead to pay them to the lodge secretary at home or in the pub.[139] Accordingly it was not until well into the second half of the nineteenth century that trade unionism began to make much more than the most marginal impact on the day-to-day life of the ordinary miner.

Unfortunately it is not easy to assess accurately the impact of short-lived, informal, and often secret, early nineteenth-century organisations like the 'Scotch Cattle' of South Wales.[140] What evidence there is, suggests however that these incipient unions had only the most passing existence, let alone importance. The early 1830s were a high point of early nineteenth-century union activity. with organisations being formed in almost every coalfield. But they rarely lasted more than a year or two. In March 1830, South Lancashire colliers met at Bolton to form a Friendly Society of Coal Mining; a strike ensued, the men were defeated and 'In August 1831, the Friendly Society of Coal Mining was no more.'[141] Even the most successful unions were little longer lived. Tommy Hepburn formed his Coal Miners' Friendly Society in 1830. Most Northumberland and Durham miners joined and within a year the society had paid over £13,000 in sick and funeral benefits and more than £19,000 in unemployment pay to those on strike. In 1831 the men came out on strike, winning wage increases of about 30 per cent, reducing boys' hours from fourteen to twelve a day and improving the terms of the bond. But within a year the union lay in ruins and 'The great man who had led the miners during their struggles in 1831 and 1832, now very shabbily clad, no one to converse with, broken down in spirits, proscribed and hunted, had to go and beg at the Felling for employment.'[142] Then there was the Miners' Association of Great Britain and Ireland. Formed in 1842, by 1843 it had

50,000 members (about a third of the country's miners),
and 'for the major portion of 1844 . . . achieved the miracle
of being a genuinely national trade union.' It broke the bond
in Durham and Northumberland and reduced hours of work
in the East Midlands. But by 1848, just six years after its
formation, the Miners' Association had ceased to exist as an
effective organisation.[143]

Not that the members of these pioneer mining unions
laboured entirely in vain. The tradition of union activity was
established. As Tommy Hepburn pointed out in almost his
last recorded speech:

> If we have not been successful, at least we, as a body of
> miners, have been able to bring our grievances before the
> public; and the time will come when the golden chain
> which binds the tyrants together, will be snapped, when
> men will be properly organised, when coal owners will only
> be like ordinary men, and will have to sigh for the days
> gone by. It only needs time to bring this about.[144]

Time was one thing that was on the miners' side. The conditions necessary for the growth of stable trade unionism
were much the same as those needed for the development of
other kinds of self-help activity. The greater the stability of
the population, the greater its homogeneity and the higher its
earnings, the more likely were firmly established trade unions
to emerge.[145]

Slowly these conditions were beginning to come about.
Slowly too the unions began to woo — or at least to accept —
that forgotten two-thirds of the miners who did not work at
the coal face. This new acceptance derived in large part from
the patent dissatisfaction of the oncost workers. From the
1870s onwards the deputies and craftsmen began to develop
their own small, specialist unions like the Durham Colliery
Enginemen's and Boiler Minders' Mutual Aid Association
which was started in 1872, and the South Wales and Monmouth Winding Enginemen Association and Provident Trade
Union which was established in 1895.[146] Surface workers too
were unhappy with the major unions. So by 1873 there was a
South Yorkshire Topmen's Association in existence and
twenty years later there were surface unions based in both

Cheshire and in South Wales where the 'Hauliers and Wagemen of South Wales and Monmouth' had formed their own union to represent their interests against the large, hewer-dominated associations.[147] By the start of the twentieth century about half the surface workers in Yorkshire and Derbyshire belonged, not to the Yorkshire or Derbyshire Miners' Associations, but to the National Federation of Colliery Surface Workers.[148] Some surfacemen were even beginning to turn to unions from outside the industry. In the Rhymney district of South Wales they joined the Gas Workers' Union and in Derbyshire the National Amalgamated Union of Labour was recruiting surface workers until it had about 2,000 in 1912.[149] The result of all this was that by the turn of the century the major mining unions were concerning themselves more and more with the interests of the whole of the colliery labour force. Many unions became less exclusive. The Denbighshire and Flintshire Miners' Federation accepted surface men from 1894 onwards.[150] In 1900 the West Bromwich and District Miners' Association decided to admit the previously proscribed engine men and surface workers and in 1907 the rules of the Lanarkshire Miners' County Union were 'altered to make it easier for those employed at the pits aboveground to join'.[151] Some unions became more receptive to the needs of the oncost men. The Derbyshire Miners' Association had always practised an open door membership policy but the success of the National Amalgamated Union of Labour forced the association to pay more positive attention to the position of the surface workers. Indeed it is significant that in 1912 the annual conference of the Miners' Federation of Great Britain decided to take up the question of banksmen's wages and demanded a minimum wage of 5s. (25 pence) a day for all adult surface workers.[152]

Slowly then union membership began to pick up. But it remained low well into the second half of the nineteenth century. In the early 1860s fewer than a tenth of all miners were unionised and even the optimistic Alexander MacDonald could bring himself to claim only a fifth of Scottish miners as union supporters.[153] As in every up-swing of the trade cycle, so the boom of the early 1870s saw a sharp rise in union activity. But it is important not to exaggerate what was still very definitely a minority interest. While it is true that over half the men in Yorkshire, Northumberland and Durham and the small Gloucestershire-Somerset coalfield did join their local union, membership elsewhere was far lower. Well under half the labour force was enrolled in the Midlands, the West Midlands, Scotland and in Lancashire.[154] So even in this unprecedented boom only about 40 per cent of the men actually joined their local association. Indeed the vigorous efforts made to eradicate non-unionism during this period testify only too vividly to the extent of the problem. Nearly 500 miners employed by Merry and Cunninghame at Beith in Ayrshire refused to work with twenty non-unionists whom the manager refused to dismiss.[155] At Plymouth near Merthyr Tydfil the men held a 'show card' day to stop five non-unionists from working and the ringleaders were summoned for intimidation and breach of contract.[156]

Despite such efforts, membership slumped dramatically during the depression of the late 1870s. Unionism was most soundly established in Northumberland and Durham but even here membership fell from 55,000 in 1875 to 40,000 five years later.[157] In other parts of the country the decline in union support was often catastrophic. The Bulgill lodge of the West Cumberland Miners' Association had once had 300 members; by 1878 it was down to eleven and it had to close.[158] In the Midlands, the Derbyshire and Nottinghamshire Miners' Association was able to boast a membership of 5,000 in 1874. Four years later it was 'virtually extinct'.[159] In Scotland, membership of the Miners' Society of Mid and East Lothian fell steadily from over 2,000 in 1872 until the union had to be dissolved at the end of the decade.[160] Union membership remained low, lagging well behind the support being given to other agencies of self-help. In 1885 only about a fifth of the country's miners were unionised — compared to nearly a third who belonged to permanent relief funds.[161] Again there were pronounced regional variations. But it is interesting that districts like Northumberland and Durham which were most strongly unionist were also the most loyal supporters of the permanent relief funds. The Durham Miners' Association and the Northumberland Miners' Mutual Confident Association between them had just over 50,000 paid up members; the

Northumberland and Durham Miners' Permanent Relief Fund had over 86,000.[162]

It was not until the final decade of the century that miners joined the unions in sufficient numbers either to outstrip these other self-help societies or to make a significant day-to-day impact on the lives of miners and their families. Membership was still low in Scotland, Lancashire and South Wales but the two largest North-Eastern unions contained over half the region's miners; the Midland Federation two-thirds of those in Staffordshire, Worcestershire, Somerset and Gloucestershire and the Yorkshire Miners' Association about three-quarters of those working in that county. In all, nearly half the miners in Britain now belonged to one trade union or another.[163] Thereafter unionisation proceeded apace so that by the turn of the century three-quarters of South Wales miners had joined the South Wales Miners' Federation and two out of every three miners employed in the British coal industry were fully paid up members.[164] Weak points remained of course. The South Wales Federation battled manfully against non-unionism. At the Lady Windsor lodge 'the correct number of defaulters, when they were excessive, was rarely given, so that they might not know their "strength".' More positive action was also taken. 'A sight to gladden a Lodge official's heart happened one summer morning, when . . . a long "goot", three or four deep, extended from the pit as far as the lamproom, every man flaunting his yellow show-card.'[165]

We still know little enough about the causes, incidence and impact of such non-unionism. We know enough, however, to see how seriously misleading it is to think of the history of the miners as if it were synonymous with the history of their unions. It was only from the 1890s onwards that even half the labour force was enrolled year in and year out, and only by the very end of the period that four out of every five miners at work in Britain belonged to one or other of the many mining unions.[166] But this growing membership was matched by a commensurate expansion of ambition and activity so that from the end of the nineteenth century the unions were able to exercise a really decisive influence on the life of the individual mining family. They made prodigious efforts to improve the assistance available to their members when they were ill, injured, unemployed, on strike or just too old to work. Union leaders stood at Board of Guardian elections and tried, with scant success it must be admitted, to push through more generous policies. In the early summer of 1893, James Haslam, secretary of the Derbyshire Miners' Association, urged his fellow members of the Chesterfield Board of Guardians to grant outdoor relief to the families of miners on short time. 'It is not desirable', he argued, 'that poor people should break up their homes to go into a work-house . . .' Something should be done 'to remedy the grinding unfairness of the parochial law'. But nothing was done. Despite the fact that about 9,000 local miners were working a two-day week, another member of the Chesterfield Board denied that there was any 'exceptional distress in the district'.[167]

From the last quarter of the century the unions, often through their local agents, worked to secure some legal redress for injured miners. The Durham Miners' Association decided as early as 1870 that in every case where a manager's negligence resulted in the death of a member, the matter should be brought before a meeting of delegates from every colliery which would then attempt to recover compensation. But it was not until after the passing of the Employers' Liability Act in 1880 that it became common for unions to meet the cost of claiming legal redress. Despite its poverty, the Derbyshire Miners' Association helped William Flint to pursue a claim for damages against the Blackwell Colliery Company; the Durham Miners' Association set up a Defence Fund to meet legal costs and the Lancashire and Cheshire Miners' Federation supported several claims for compensation.[168] Some lodge secretaries rather let their zeal for justice run away with them. In 1910 John Chapman, secretary of the Kiveton Park branch of the Yorkshire Miners' Association, was pretending to be a solicitor when writing letters on behalf of a Mrs Asher who was trying to get compensation under the Workmen's Compensation Act of 1897.[169]

Unionists were little more successful in gaining a voice on the public appeals raised to relieve distress in mining areas.[170] So like it or not — and many apparently did — the unions were forced to help themselves. Branches did what they could on an *ad hoc* basis. In the winter of 1873-4, for instance, the Wearmouth lodge of the Durham Miners' Association distributed a kell of coals among the poor while the East Hetton lodge paid out £17 as Christmas and New Year gifts to local widows, orphans and paupers.[171] But more and more unionists were driven to the belief that the best way for miners to survive periodic crises of unemployment, sickness and old age was to insure against them. Members were encouraged to join friendly societies when these did not compete directly with the unions' own funds. Union leaders opposed the establishment of permanent relief funds only so long as they were organising their own, rival insurance schemes. But as union funds were abandoned during the late 1870s, so support was transferred to the permanent funds. The agent of the North Staffordshire miners admitted candidly that 'So long as the widows and orphans fund was in existence in connection with his own trade society he could not advocate the claims of the Permanent Relief Society, but now that his fund had ceased to exist, owing to the continued depression of trade, he should do all in his power for the interests of the Relief Society.'[172]

Accordingly during the second half of the nineteenth century the mining unions came to provide their members with a whole range of insurance schemes. Many allowed subscribers to insure against unemployment. Earlier chapters delineated the heavy strain placed upon family life in the coalfields, as elsewhere, by the cyclical pattern of boom followed by slump, an abundance of work followed by a shortage. So in 1875 the Northumberland Miners' Mutual Confident Association decided 'That when a colliery is laid idle for repairs, men living in rented houses to be allowed their rent by our association.'[173] Within a few years a Stoppage Fund was started to help men whose pits had been idle for seven successive working days and by 1886 the scheme had been modified to take account of short-time working. By this Special Relief Scheme, members in Northumberland were entitled to 1s. 6d. (7 pence) for every day over ten days that they had been idle during the previous month.[174] Other county unions reacted to unemployment in similar fashion. In 1895 the Point of Ayr lodge of the Denbighshire and Flintshire Miners' Federation decided to pay members who had been out of work for a fortnight the sum of 5s. (25 pence) a week from lodge funds.[175] In 1892-3, and again from 1895, the lodges of the Derbyshire Miners. Association paid the unemployed at the rate of 7s. (35 pence) a week for married men and 5s. (25 pence) for the single. Low as the payments were, they relieved a vast amount of distress. During just three months in the summer of 1905 the Association distributed over £40,000 in unemployment benefit.[176] Another way of dealing with unemployment was by encouraging emigration to more prosperous parts of the world. In the late 1870s the management committee of the North Wales Miners' Association resolved to vote £7 to every member going to the United States and twice that sum to anyone emigrating to Australia or New Zealand.[177] The Northumberland miners adapted their Stoppage Fund to pay 'shifting money' (consisting of £2 plus 12s. 6d. (63 pence) for each child under twelve) to any member of a year's standing who decided to leave the country.[178]

Superannuation schemes though not common were, however, started. Small unions like the Northern United Enginemen's Association helped their elderly members as did larger organisations like the Mid and East Lothian Miners' Association, the Leicestershire Miners' Association and the South Yorkshire Miners' Association, the latter paying from 5 − 9s. (25 − 45 pence) a week according to the length of membership.[179] Other old-age pensions were begun later. It was decided in 1906 that every miner who had been in the Derbyshire Miners' Association for fifteen years was to receive, without any further contribution, a pension of 5s. (25 pence) a week for life at the age of sixty if he was unable to work.[180]

Far more common and far more important were the benefits provided for sick and injured miners and their families. The odd union ran its own rest-home. The Lanarkshire miners voted £30 towards a 'Miners' Convalescent Home at Saltcoats' and the Derbyshire Miners' Association inherited a convalescent home at Skegness which by 1914 was dealing with 479 patients.[181] But all over the British Isles the usual way of helping was by offering members the facilities for insuring against illness and accident.[182] Nearly two-thirds of English mining unions organised some form of accident insurance during the second half of the nineteenth century. A fifth ran widow and orphan funds, two-fifths offered injury insurance facilities and a half organised their own funeral funds. Typical of the latter was the funeral branch of the Farnworth and Kearsley Miners' Permanent Benefit Society in Lancashire which paid £5 to the dependents of any subscriber killed during the course of his work. In 1875 a quarter, and from 1890 a third, of all English miners were members of these trade union funeral funds. Here again the branches were important. It was not just because it was at branch level that subscriptions were collected and benefits paid. Many branches organised their own individual funds. In the 1870s, for example, there were local sickness and accident schemes in operation at the Strafford, Kilnhurst and Wharncliffe Silkstone lodges of the South Yorkshire Miners' Association.

Important though these sickness and accident funds undoubtedly were, they never rivalled the benefits paid to members who were locked out or on strike.[183] Industrial disputes were common and corrosive. The violence, unpredictability and danger of underground work, the high labour costs of the industry, the instability of earnings and the constant wage negotiations all led to a high strike rate.[184] There were major disputes in every coalfield: in South Wales in 1816, 1822, 1830, 1871, 1873, 1875 and 1898; in the West Midlands in 1800, 1816-17. 1821-2, 1826, 1830, 1842 and 1884; in Northumberland and Durham in 1810, 1816, 1825, 1831, 1844, 1863-4, 1879, 1887 and 1892.[185] Some pits went years without a strike; others had their own local conflicts almost every year. At West Cramlington in Northumberland, for example, there were twenty-three strikes in the twenty-two years between 1841 and 1863.[186]

The suffering, despair, misery and poverty caused by such bad industrial relations is well known and not to be underestimated. Family was set against family. The funerals of the relatives of those men who worked during the Aberdare strike of 1850 were 'saluted . . . with the most heartless yells and laughter and by the discordant sounds from the beating of frying pans, kettles etc.'.[187] Children went hungry and the savings of years soon disappeared. A local clergyman described the Christmas endured by the children of the miners evicted during the Denaby bag-muck strike of 1902—3:

> My band of willing workers were all impressed with the same fact, namely that the children were desperately hungry. They arrived at this conclusion through the eagerness of the little ones for plain bread. We thought that it being the festive season they would plunge into sweet cake, but the children asked for bread and butter.[188]

Industrial disputes were a major cause of coalfield poverty and it was by the provision of strike and victimisation benefit that the mining unions, both big and small, made their greatest contribution towards easing the financial needs of their members. There is space here for only a few examples. South Yorkshire strike pay during the 1870s was the same as that for sickness: 8s. (40 pence) a week, with 5s. (25 pence) extra if married and a shilling (5 pence) more for each child under the age of twelve.[189] By the end of the century the North Wales Miners' Association and the Derbyshire Miners' Association were both making strike and victimisation payments of 10s. (50 pence) a week.[190] It is easy to imagine how such payments added up. Tommy Hepburn's Coal Miners' Friendly Society spent over £19,000 on strike pay in less than a year. Payments to 1,500 men in Derbyshire and 1,300 in Lancashire were costing £1,500 a week in 1867; the 1893 Lock-Out Relief Fund cost the Miners' Federation of Great Britain £75,000; and the Lanarkshire County Union spent over £23,396 on strike and lock-out pay in 1908, a year in which there was no major dispute.[191] It is perhaps less easy to imagine what the payments meant in human terms.

> My father had become a member of the committee of the local Branch of the Union and I attended every Branch meeting with him. As he could not read, it was my job to read out to him the minutes of Yorkshire Miners' Delegate Meetings, which were attended by a delegate from each branch . . . Strike and lock-outs and hardship and poverty were, of course, vital influences in my life at this time. But, looking back, I think that the simple sort of entry that recurred time and again in those minutes I read to my father

was a bigger influence. For example: 'Mr X at Y Colliery.
That Mr X be paid two weeks' victim pay, seeing that he
has been taken undue advantage of by the Management.'

This meant that Mr X had been dismissed; that the full
circumstances of the dismissal had been investigated by
officials of the Union and that the deleagates (*sic*) were
satisfied that Mr X had been dismissed without good cause.
So Mr X was awarded two weeks' victim pay at the rate of
nine shillings a week. But he had lost his job . . .[192]

It is important not to exaggerate the impact of industrial
disputes, however, for just as historians have been bemused
by the multiplicity of price lists and piece rates in the industry,
so some have been seduced by the seemingly endless proces-
sion of strikes and lock-outs. But to discuss the history of
the miner in terms of industrial disputes is as misleading as
describing it solely in terms of trade union membership and
bargaining strength. The burden of strikes, like everything
else, fell particularly heavily on the women. But, as D. H.
Lawrence recognised, it might not be so bad for the men
themselves.

> Oh, strike sets the men up, it does. Nothing have they to
> do but guzzle and gallivant to Nottingham. Their wives'll
> keep them, oh yes. So long as they get something to eat at
> home, what more do they want! What more *should* they
> want, pri thee? Nothing. Let the women and children starve
> and scrape, but fill the man's belly, and let him have his
> fling. My sirs, indeed, I think so! Let the tradesmen go —
> what do they matter! Let rent go. Let children get what
> they can catch. Only the man will see *he's* all right.[193]

This is an exaggerated view, but there is more than a little
truth in it. During strikes at one Rhondda colliery, 'Most
of the rank and file . . . seemed to enjoy themselves in spite
of short rations. The stoppages generally took place in summer, and almost without exception were accompanied by
extremely fine weather. Young lads went bathing; older men
took to their allotments.'[194] It must also be remembered —
though this is nearly always overlooked — that in many ways
industrial disputes were less disruptive of coalfield life than

the less well-known accidents and diseases. Even at the end of the nineteenth century, when accident rates were on the decline, more working days were lost from industrial accidents than as a result of strikes and lock-outs. In 1896 over one and a half million working days were lost after non-fatal injuries in the English coalfields whereas only just over a million days were lost because of all disputes in mining and quarrying throughout the whole of Great Britain and Ireland. Similarly in 1897, 867,000 working days were lost in industrial disputes, but 1,650,000 days were lost as the direct result of injuries received in the pit.[195]

If one weakness in the study of mining trade unionism has been the undue emphasis placed on the miners' strike record, then another has surely been the excessive attention paid to the plethora of speeches, amendments and instructions emanating from the central bodies of the unions. As unions grew in size, so their permanent officials tended to become more and more isolated from the rank and file.[196] One South Wales miner wondered 'why our leaders always hold the conferences at pleasant places like Blackpool and Margate? I have heard men ask one another that question many times. Why not hold them in the Rhondda Valley, or at Landore or Llansamlefar or on the Tyneside?' He answered his own question: 'Every year our leaders spend away from the very heart of the industry makes them feel more contented with the conditions of those they represent.'[197] He was not alone in his disquiet. Writing after the General Strike, a miner remarked that

> There has always been a smouldering fire of antagonism towards the leaders among the miners . . . The miner's job is hard, dirty and disagreeable. This gives him a permanent grouse at the world . . . when the ex-miner turned Labour leader.and fat and heavy for lack of exercise drives up in a motor to the 'mornin'' meeting at some pit, the assembled colliers, shivering at the pit gate in the chill of the morning, never fail to remark unflatteringly on his thick well-cut coat and his embonpoint. The contrast is too great to be endured.[198]

With the rank and file increasingly alienated from its leadership, with mining villages sometimes physically and often socially isolated, and with conditions everywhere varying so much from district and from pit to pit, it is no surprise to find that the local union branch came to assume a leading role. Not that the branches were necessarily models of democracy and fairness. Of the Nostell branch of the West Yorkshire Miners' Association it was said, for example, that 'Some men can be heard at a meeting even if they talk about the minister of the parish, but if a man who does not belong to the clique gets up to make a complaint, or an enquiry he is treated with scorn.'[199] Of course, admitted another Yorkshire miner, 'In the old established pits ... ties of long comradeship and family relationships naturally played a big part when the Branch officers are elected.'[200] But whatever the failings of the local branch, it was closer, far closer, to its membership than the central leadership could ever hope to get. To most miners it was the branch secretary, rather than the nationally-known general secretary, who represented the growing power and influence of the union. It was the branch secretary who collected subscriptions, saw the management about working conditions, haggled about the price list, saw to repairs to colliery houses, doled out the strike pay and dealt with the claims arising from accidents.

The branches ran their own social functions, modelled on the annual demonstrations and galas organised by their parent associations. Thus the proceeds of the tea provided at the annual gathering of the Seaton Delaval branch of the Northumberland Miners' Mutual Confident Association in 1873 went to the local brass band which played between the 'purring, jumping, pole-leaping, old aunt Sally, and other sports [which] were indulged in by a numerous company'.[201] More important, the branches tried to improve the range and quality of miners' housing. They put particular pressure on the owners of tied property to ensure that the available housing was allocated fairly among their members. So when a joiner was put into one of the houses being built by the Consett Iron Company in the 1870s the men struck work until they were promised that the newcomer would be made to leave.[202] The branches also fought to improve water supplies. The Northumberland Miners' Mutual Confident Association took up the case of eighty men at the Backworth collieries and 150 more at Dinnington who were threatened with the sack unless they agreed to pay for the water supplied by the local water company.[203] And the Association decided to pay the expenses of the Burradon colliery men who went on to fight their employers through the courts.[204] The branches of the North-Eastern unions also gave their support to the Northumberland and the Durham Aged Miners' Home Associations which were started in the 1890s. The first home, 'a large hall and two acres of ground, situated near Boldon Colliery . . . At the request of the Boldon workmen . . . were given to them for the use of their old people on their promise to be responsible for the repairs and rent.'[205] Like their leaders, local union officials were quick to see that owning property was one way to remove the fear of victimisation. Accordingly in Durham every lodge built cottages for its, chief officials as soon as it could possibly afford to.[206] Nor were unionists slow to point out to the ordinary member the advantages of owning his own home. Hector McNeill, the Larkhall district agent, urged his members to form co-operative building societies 'so that they might be independent of their employers for homes'.[207] Ben Pickard argued that the home-owner 'will not only get social freedom, but also political power. There will be no rent days, and nothing to cause him to fear the unruly landlord's voice giving him notice to quit the house which, perhaps, he has been at much trouble and expense to make comfortable.'[208]

Centrally and locally, all unions put their greatest efforts into trying to improve — or at least to maintain — working conditions and take-home pay. There was the perpetual problem of abnormal places, those wet, thin, faulted, hard, gassy or distant seams, in which the piece-worker, through no fault of his own, was unable to earn a living wage. Naturally it was a great temptation for management to fill abnormal places with militants and other types of trouble-maker. To this the unions responded in several ways. In Lancashire the Amalgamated Association of Miners insisted on taking the tubs from the hewers in strict rotation, so that they all worked at the pace of the slowest.[209] In Northumberland and Durham the unions insisted on the maintenance of the cavilling system, whereby all working places, good and bad, were allocated by a quarterly ballot known as the cavil. 'Let the cavillings remain,' insisted the Durham Miners' Association in 1882, 'Even with them, we have an amount of favouritism amongst overmen. Without them, our leading men might call for heaven's aid.'[210] So when, for example, men were sacked at a Northumberland colliery in 1904 for refusing to work in a wet place to which another miner had been cavilled, the executive committee of the Northumberland Miners' Mutual Confident Association decided to pay them all the sacrificed allowance.[211] The unions recognised that somebody had to work in the abnormal places and efforts were made to help those unfortunates by asking for instance 'That men be allowed to ride as soon as they come to the shaft, and especially when they are wet.'[212]

If the unions were unable to improve conditions underground, they could at least seek to ensure that their members spent as little time as possible in them. Alongside the (ultimately) successful campaign to reduce the length of the working day, the unions worked locally to ensure that their members worked fewer and shorter shifts. They demanded polling day holidays; they fought for a Saturday half-day and they refused to work on Sunday and on other shifts as part of their policy of output restriction.[213] At the Eddleswood colliery near Hamilton it was union policy in 1880 not to work on Thursdays. One week seventy men defied the union and went to work. But the next day sixty-nine of the seventy changed their minds, stayed off work and went on strike against the one man who did not come out.[214] The unions pressed for proper summer holidays. In 1905 fourteen of the seventeen districts in Mid and East Lothian asked for a week's stop-work in the summer to coincide with the Edinburgh trades' holidays. The owners refused but eventually, a compromise was reached. The men got three days starting from the Lothian Miners' annual gala which was held in Galashiels at the end of July.[215]

However successful or otherwise the unions may have been in improving working conditions and hours of work, they sought constantly to exert a direct influence over the level of wages. This they were able to do by the last quarter of the nineteenth century. At county level wages were regulated by sliding scales, joint committees and joint conciliation boards, all of which, whatever their faults, did contain representatives from the union side.[216] The unions were just as active at local level. It was made clear in the earlier discussion of the industry's wages structure that by the end of the century the union branch was playing a vital role in this process.

> The workmen of a colliery . . . form themselves usually into a "lodge" or branch of the Miners' Federation of the coalfield, negotiate with the owner of the colliery a series of piece rates or "prices," to be paid for work done . . . The negotiation of a price list is indeed a most dramatic event in the life of a mining town, the prosperity and character of the town hanging in the balance on the "give and take" of a 1d. or ½d. per ton.[217]

Here is a South Yorkshire miner's description of a branch committee negotiating the price list at Barnburgh Main, a new colliery where 'most of the work . . . was still around the pit-bottom, as development outward had not yet begun.'

> The committee's first job in a new mine is to negotiate with the management a price-list. The price-list is a small booklet which sets out how much is to be paid for every kind of work in the mine. As far as the colliers are concerned it is a very complex document, for the money paid varies according to the nature and difficulty of the work involved. The list is therefore a vital document. Once established, it can only be changed with difficulty, and complete revision is not lightly or frequently undertaken . . . The prices that have to be fixed are concerned with every phase and type of operation in the mine. Account is taken of the height of the seams to be worked, of the extra difficulty of working in wet places or where a fault runs through the seam, and so on. None of this can be left to be worked out at a later stage, though occasional amendments are possible. The men rightly want to know what each job is going to be worth before they start it and bad pricing will only lead to endless trouble in which neither side will be the gainer. . . . Cubic and lineal measurements

had to be worked out and then explained to inexperienced colleagues so that they could make the best use of them in argument with the trained officials representing the management. At this stage also the colliery officials have an even greater advantage than their negotiating skill and experience normally gives them. The situation is this: the colliers, the men who actually hew the coal from the seam, are all on day rates, at and around the pit-bottom. Then work starts on the seam. Immediately the colliers go on to piece rates, that is when the price-list is settled. As the coal-faces are still near the pit-bottom, the colliers lose very little time travelling between the pit-bottom and the working faces. Consequently, whatever the rates temporarily in force before the list is finalised and operating, the colliers are earning good money. They spend practically their entire shift getting coal and do very well. In these circumstances they tend not to think about the future. Unless firmly warned and led they are liable to accept rates which may be all very well while they have no distance to travel to their working places, but which will be mean indeed when the work of the pit has expanded further away from the lift-bottom. Naturally the management play on this short-sightedness as much as they can. The response of the Branch officials is to prolong negotiations for as long as they dare, to give time for the realities of the situation to be borne in upon the men. In the case of Barnburgh Main the negotiations were protracted over many months. Eventually the stage was reached when the Union's permanent officials took a hand and final terms were recommended to the men. These were accepted in a ballot of the whole Branch.

As this passage makes clear, it was difficult to alter the price list once it was fixed. It became valuable to one side or the other.[218] But from the 1890s onwards it was the duty of the local committee to ensure that awards made at county level by the conciliation boards were applied correctly in each individual price list.[219] And it had always been the committee's duty to bargain for extra payment when working conditions changed for the worse or when a member was required to undertake work which for some reason was not

shown on the price list. Thus at the end of the century workmen employed in the 2' 9" seam at the National collieries,
Wattstown, Pontypridd began to claim payment for the
deposit that was found to be overlying the top coal.[220] The
introduction of new working methods always afforded the
opportunity for renegotiation. The Clay Cross miners wanted
new arrangements to take account of the extra dust caused
by mechanisation.[221] The men always argued that the safety
lamp gave off such poor light that their earnings were seriously
reduced. 'It was ridiculous', claimed the delegate from one
Derbyshire lodge, 'to think that the men could earn as much
with safety lamps as they could with naked light.'[222] At some
collieries in Northumberland and Derbyshire — and probably
elsewhere — the men were able to win small concessions of
perhaps a penny or twopence a ton.[223] So the miners' hostility
to new methods was not the simple Luddism that it sometimes seems. There was invariably a strong financial motive
for their conservative responses. Another, less direct, way of
seeking to upset the price list was to look to the men's perquisites. Efforts were made to improve the supply of concessionary coal.[224] Then, in 1905, the Northumberland miners
claimed that young unmarried men who were acting as heads
of their households should receive the same rent allowances
as were being afforded to married men — in other words that
they should get a pay rise.[225]

Trade unionism, like every other self-help movement in the
coalfields, was nurtured by the same three developments in
family life: greater stability, fewer children and higher wages.
The result was that by the end of the century the number of
accident insurance policies alone held by miners with local
friendly societies, with the affiliated orders, the collecting
societies, the Prudential, the trade unions and the permanent
relief funds exceeded by a fifth the total number of miners who
were at work in the industry.[226] Another result of the changes
in family life was that by the end of the century the mining
unions were able to exercise a decisive, day-to-day influence
over the lives of ordinary miners and their families. Not that
the unions, any more than other self-help agencies, were able
to solve the social problems of the coalfields. But it is apparent
that the traditional picture of the thriftless miner is a carica-

ture. Even if trade unionism often brought few immediate and tangible benefits, it did promote self-awareness and the social satisfaction of combined action.

The union lodge, like the co-op and friendly society, both reflected and encouraged the miners' foresight and self-respect. In Durham, remembers Sid Chaplin,

> each village was, in fact, a sort of self constructed, a do-it-yourself counter environment, you might dub it. The people had built it themselves. There was everything there, in the pit village. The Lodge was there to look after the interest of the men at the pit, and they had built up a kind of equipoise with the coal owners. They had fought them to a standstill. And this remained too until roughly 1926, and then the break-up came.
>
> But in addition to this you had a sort of complete welfare system. You had your Sons of Temperance, and you had your Rechabites and you had sick clubs. You had your Hearts of Oak, which was traditionally for the mechanics and you had your Workmen's Club, which also had its social and educational side. Very much stronger than it is now. And you had your aged miners' homes, and you had your hospital scheme, which was run through the union. And there was the Lodge secretary with his galluses (that's his braces) hanging down his side. And his door was never shut, his door was always open. Jug Goundry, as rough a diamond as you could ever wish to meet, blind drunk every Friday or Saturday night and pretty often on a Sunday night as well, but an encyclopaedia of social welfare. If you were in trouble, if you had an accident down pit, if you were negotiating a new agreement, if you wanted an aged miner's home, if you had trouble with your coal — that is your concessionary coal — if you wanted some roof repairs doing on your colliery house, if your lad had to go to hospital and you needed the ambulance, well you went to Jug.[227]

Conclusion

This book has attempted to bring to the history of coal-mining some of the approaches currently being adopted by other historians of the working-class. Labour history has become less hagiographic and less narrowly institutional in recent years; slowly the history of industrial labour is being rewritten from the standpoint of the rank-and-file workers rather than from the vantage point of the union head office. Labour history is attempting to rescue working people, in E. P. Thompson's celebrated phrase, 'from the enormous condescension of posterity'.[1] Coal-mining historiography has not been unaffected by this growing interest in the social history of the working-class. Efforts have been made to preserve the records of mining labour and a number of historians are turning towards the detailed study of social conditions in the British coalfields. But a great deal remains to be done. Of the fourteen postgraduate theses and dissertations on miners' history that were completed between 1974 and 1977, four were primarily concerned with trade union developments, two dealt with unofficial strikes and syndicalism, three with industrial relations and the wages structure, and one with the Forest of Dean miners' riot of June 1831. Only three of the fourteen addressed themselves directly to the social history of the miners.[2]

It can come as no surprise, however, that nineteenth and early twentieth-century miners were never the broadly homogeneous group of popular imagination. Indeed the regional perspective of the standard trade union histories should warn against any blanket, national generalisations. The colliery population was large, scattered and growing rapidly. Conditions varied geologically, industrially, economically and socially both between and within coalfields. Whatever the value of the concept of a 'labour aristocracy', and whether or not any miners should be included within it, there were marked variations within the ranks of the colliery population.[3] Another recurring theme has been the way in which miners and their families shared in the gradual improvements in working-class standards of living during the second half of the nineteenth century. Real wages rose, families became smaller, the range of domestic and leisure amenities expanded and the rudiments were laid of a non-parochial system of social security. These were impressive advances but they make it more difficult than ever to generalise with any confidence about 'the nineteenth and early twentieth-century miner'.

Even taking this warning into account, there is at least one generalisation about the mining community that does follow from the evidence presented in this book: miners and their families were far more thrifty and responsible than is generally allowed. They turned to self-help, both individual and collective, as the best way of surviving, let alone progressing, within the existing economic, social and political framework.[4] It was an unexciting development, but most important. The nineteenth and early twentieth-century coal miner was not the man that he appears in the existing literature.

Statistical Appendix

1. Coal Production (millions of tons)[1]

District	1800	1840	1864	1880	1900	1908	1913
North-East	2.5	8.0	23.0	34.9	46.3	53.9	56.4
Lancashire and Cheshire	?	?	11.7	19.8	25.5	24.5	24.6
Yorkshire	?	?	9.3	17.5	28.3	34.9	43.7
Midlands	?	?	7.3	14.5	27.9	35.2	38.8
South Wales	1.5	4.5	11.7	21.2	39.3	50.7	56.8
East Scotland	1.2	?	6.4	4.7	11.2	17.2	17.5
West Scotland	.8	?	6.3	13.6	22.0	21.7	25.0
West Midlands	1.2	?	15.4	14.6	14.8	15.3	15.8
Cumberland	?	?	1.0	1.7	2.0	2.1	2.2
South-West	?	?	1.8	2.0	2.6	2.7	2.9
North Wales	?	?	1.8	2.4	3.1	3.4	3.5
Kent	—	—	—	—	—	—	.1
Northern Ireland	?	?	?	.1	.1	.1	.1
Great Britain	10.0	30.0	95.6	147.0	225.2	261.7	287.4

1. The figures for 1800 and 1840 are estimates. The national totals are not always the same as the sum of the regions because of occasional mining in for example Devon and Lincolnshire.

Source: Hair, 'Social History'; *Reports of H. M. Inspectors of Mines.*

2. Numbers Employed in the Coal Industry (thousands)[1]

District	1800	1840	1864	1880	1900	1908	1913
North-East	13.5	27.0	53.5	94.8	152.6	194.1	226.8
Lancashire and Cheshire	5.2	20.7	41.8	62.7	88.0	100.6	109.0
Yorkshire	3.5	13.8	34.5	60.4	107.9	143.9	161.2
Midlands	2.1	8.3	26.6	49.3	90.3	116.6	130.5
South Wales	1.5	15.2	45.9	69.0	147.7	201.8	233.1
East Scotland	1.8	7.3	21.2	35.5	38.1	59.3	68.5
West Scotland	3.2	12.7	18.1	41.0	65.7	72.1	79.1
West Midlands	6.0	23.6	46.1	43.9	53.8	58.8	66.8
Cumberland	.5	2.3	4.0	5.9	8.5	9.6	11.0
South-West	1.8	7.1	9.9	11.1	13.1	14.7	15.5
North Wales	1.2	4.9	6.1	10.2	12.6	14.9	15.9
Kent	—	—	—	—	.1	.3	1.1
Northern Ireland	?	?	?	1.1	1.0	.9	.8
Great Britain	40.3	142.9	307.7	484.9	780.1	987.8	1127.9

1. The figures for 1800 and 1840 are estimates. The national totals are not always the same as the sums of the regions because of occasional mining in for example Devon and Lincolnshire.

Source: Hair, 'Social History', 6, 15–16, 27a–b; *Reports of H. M. Inspectors of Mines.*

3. Numbers Killed and Injured in the English Coal Industry, 1865–95[1]

(a) Killed

District	1865	1875	1885	1895
North-East	179	149	169	150
Lancashire	157	169	303	122
Yorkshire	55	267	85	91
Midlands	82	65	73	61
West Midlands	143	164	71	156
South-West	18	24	17	17
England	634	838	718	597

(b) Injured

District	1865	1875	1885	1895
North-East	17,900	14,900	11,400	15,000
Lancashire	14,400	16,200	12,500	11,700
Yorkshire	5,500	11,100	8,500	9,100
Midlands	7,400	6,500	7,300	5,400
West Midlands	13,200	11,000	5,400	7,900
South-West	1,800	2,400	1,700	1,000
England	60,200	62,100	46,800	50,100

1. The method used to estimate the number of injuries is described in Chapter 2.

Source: Benson, 'Compensation', Tables IV, VIII.

4. Colliery Accidents Claiming over 100 lives

Year	Colliery and Location	Lives lost
1835	Wallsend, Northumberland	102
1856	Cymmer (Rhondda), Glamorgan	114
1857	Lundhill, Yorkshire	189
1860	Black Vein, Risca, Monmouthshire	142
1862	New Hartley, Northumberland	204
1866	Oaks, Yorkshire	361
1867	Ferndale, Glamorgan	178
1875	Swaithe Main, Yorkshire	143
1877	Blantyre, Lanarkshire	207
1878	Wood Pit, Haydock, Lancashire	189
1878	Abercarn, Monmouthshire	268
1880	Black Vein, Risca, Monmouthshire	120
1880	Seaham, Durham	164
1880	Naval Steam Coal, Penygraig, Glamorgan	101
1885	Clifton Hall, Lancashire	178
1890	Llanerch, Monmouthshire	176
1892	North's Park Slip, Tondu, Glamorgan	112
1893	Thornhill, Yorkshire	139
1894	Albion Cilfynydd, Glamorgan	290
1905	National, Wattstown, Glamorgan	119
1909	West Stanley, Durham	168
1910	Wellington, Cumberland	136
1910	Hulton, Lancashire	344
1913	Universal, Senghenydd, Glamorgan	439

Source: Duckham, *Disasters,* 202–7.

5. Membership of Trade Union Funeral Funds, the Prudential Assurance Company and Miners' Permanent Relief Funds, 1865—95

Year	Trade Union Funeral Funds	Prudential Assurance Company	Miners' Permanent Relief Funds	Total member- ship	As % of mining labour force
1865	16,000	—	7,000	23,000	1
1875	98,800	c.7,500	59,300	c.165,600	c.44
1885	56,100	c.89,000	161,300	c.306,400	c.86
1895	135,800	c.230,000	231,000	c.597,600	c.130

Source: Benson, in *EHR* XXXI/3, 417

Glossary

Anthracite: A type of coal which is difficult to light and burns with great heat, no smoke and hardly any flame.

Ash-heap: A pile of coal-ash, cinders, excrement and general domestic refuse.

Baff-day (or week): The day (or week) following pay-day.

Bait (snap): A snack.

Banksman: The workman who supervises the unloading of the tubs at the pit top.

Beat hand (knee or elbow): An inflammation caused by chaffing or irritation.

Bituminous: A generic term used to describe all coals which when heated to a moderate temperature, soften and fuse into a caked mass.

Bond: The document which bound Scottish and Northumberland and Durham miners to their employers for a year at a time.

Bord and pillar: A system of hewing whereby pillars of coal are left behind to support the roof.

Butty: A sub-contractor.

Cannel: A shiny coal that burns with a good light.

Cavilling: The allocation of working places by the periodical drawing of lots.

Chokedamp (blackdamp): The noxious gas left behind after an explosion of firedamp.

Chummins: Empty tubs.

Clarty: Muddy.

Darg: The policy of output restriction adopted in the Scottish coalfields.

Drift: A tunnel.

Driver (haulier): The worker in charge of a pit pony.

221

Enteric fever: Typhoid fever.

Filler: The workman responsible for filling coal at the face.

Firedamp: An inflammable gas given off from the coal seam.

Fireman: The workman responsible for checking the safety of the underground workings before the beginning of each shift.

Glanders: A disease caught from horses.

Glumpin': Sulking.

Hewer (getter, holer, cutter): The workman responsible for actually cutting the coal from the seam.

Hoggers: Footless, worsted stockings.

Level: A tunnel dug from the surface.

Longwall: A system of hewing whereby a number of men work along the same face.

Midden: A dunghill, manure heap or refuse tip.

Miner's asthma (pneumoconiosis): A lung disease caused by the inhalation of dust.

Miner's phthisis (black spit, silicosis): A lung disease caused by the inhalation of a stone dust found in parts of Somerset, South Wales and the East of Scotland.

Nystagmus: A disease characterised above all else by the oscillation of the eyeballs.

Oncost worker: Underground worker employed, not to cut the coal, but to service those who do.

Picking coal: Removing dirt and stone from the coal as it passes by on a moving belt.

Pitbank: The surface of the colliery.

Putter (hurrier, carter, drawer, barrowman): The workman responsible for moving wagons underground.

Riddle: The sieve used to separate the coal from the stones and dirt.

Riving and tewing: Pushing and pulling.

Screens: Surface machinery used to separate the coal into different sizes.

Self-acting fence: A fence round the top of the shaft which closes automatically when the cage descends.

Self-acting plane: A haulage system worked by gravity.

Talley: The token used to record the hewer's or putter's output.

Tippler: Surface machinery used to empty wagons of coal.

Trapper: The (generally young) worker responsible for opening and closing ventilation doors underground so as to allow the movement of men and animals without disturbing the current of air.

Truck: The payment of wages in kind.

Underviewer: The deputy manager of a colliery.

Waggonwayman: The workman responsible for an underground railway.

Winding engineman (windsman): The workman responsible for raising and lowering men, coal and equipment up and down the shaft.

Notes

Introduction
(pp. 1–5)
1. *Colliery Guardian*, 4 Mar. 1910.
2. *Wolverhampton Chronicle*, 21 Jun. 1820.
3. R. Moore, *Pit-Men, Preachers & Politics: The Effects of Methodism in a Durham Mining Community*, Cambridge 1974, 141.
4. R. Challinor, *The Lancashire and Cheshire Miners*, Newcastle 1972, 255.
5. J. Wilson, *Autobiography of Ald. John Wilson, J. P., M.P. A Record of a Strenuous Life*, Durham 1909, 30.
6. H. F. Bulman, *Coal Mining and the Coal Miner*, London 1920, 2–4.
7. *Mining Journal*, 24 Oct. 1857. Also *Colliery Guardian*, 20 Jan. 1866; *Reports of H. M. Inspectors of Mines*, L. Brough, 1856, 88; *Staffordshire Sentinel*, 12 Apr. 1890.
8. For guides to the voluminous literature see J. E. Williams, 'Labour in the Coalfields: A Critical Bibliography', *Bulletin of the Society for the Study of Labour History* 4 (Spring 1962); R. G. Neville and J. Benson, 'Labour in the Coalfields (II): A Select Critical Bibliography', *Bulletin of the Society for The Study of Labour History* 31 (Autumn 1975).
9. Williams, in *BSSLH* 4, 31.
10. R. Challinor and B. Ripley, *The Miners' Association: A Trade Union in the Age of the Chartists*, London 1968, 50.
11. R. Gregory, *The Miners and British Politics 1906–1914*, Oxford 1968, 1–2.
12. J. E. Williams, *The Derbyshire Miners: A Study in Industrial and Social History*, London 1962, 58–61.
13. K. Burgess, *The Origins of British Industrial Relations: The Nineteenth Century Experience*, London 1975, 169, 171.
14. Although I have deliberately avoided drawing examples from other countries, I am not unaware of the benefits to be derived from a comparative approach. Nor am I unaware that some foreign examples might be used to support my own case. See for example T. Zeldin, *France 1848–1945 1, Ambition, Love and Politics*, Oxford 1973, 265–8.

Chapter 1
THE GROWTH OF THE INDUSTRY
(pp. 6—27)

1. These and other figures will be found in the Statistical Appendix.
2. A. R. Griffin, *The British Coalmining Industry: Retrospect and Prospect*, Buxton 1977, 169; P. E. H. Hair, 'The Social History of British Coalminers 1800—1845', (Oxford D.Phil. thesis, 1955), 19—20.
3. Challinor, *Lancashire Miners*, 64, 80; J. H. Clapham, *An Economic History of Modern Britain: Free Trade and Steel 1850—1886*, Cambridge 1932, 121.
4. J. W. F. Rowe, *Wages in the Coal Industry*, London 1923, 1. The employment figures are contained in the Statistical Appendix.
5. B. Lewis, *Coal Mining in the Eighteenth and Nineteenth Centuries*, London 1971, 26; Griffin, *Coalmining Industry*, 9.
6. J. Evison, 'The Opening Up of the "Central" Region of the South Yorkshire Coalfield and the Development of its Townships as Colliery Communities 1875—1905', (Leeds M.Phil. thesis, 1972), 1, 116.
7. *Colliery Guardian*, 8 May 1858.
8. T. J. Raybould, *The Economic Emergence of the Black Country: A Study of the Dudley Estate*, Newton Abbot 1973, 215.
9. Evison, 'Central Region', 121; *Colliery Guardian*, 9 Apr. 1880.
10. *Colliery Guardian*, 16, 23 Feb. 1900.
11. R. W. Sturgess, *Aristocrat in Business: The Third Marquis of Londonderry as Coalowner and Portbuilder*, Durham 1975, 18.
12. *Colliery Guardian*, 13 Dec. 1872; 21 Jan. 1910.
13. *Colliery Guardian*, 16, 23 Feb. 1900. See also J. G. Worpell, 'The Iron and Coal Trades of the South Staffordshire Area Covering the Period 1843—1853', *West Midland Studies* 9 (1976), 34; A. H. John, *The Industrial Development of South Wales 1750—1850*, Cardiff 1950, 71.
14. Williams, *Derbyshire Miners*, 214; A. J. Taylor, 'Labour Productivity and Technological Innovation in the British Coal Industry, 1850—1914', *Economic History Review* XIV/1 (1961), 51—3.
15. Griffin, *Coalmining Industry*, 59—60; A. R. Griffin, *Mining in the East Midlands 1550—1947*, London 1971, 115.
16. A. J. Taylor's foreword to *Studies in the Yorkshire Coal Industry*, eds J. Benson and R. G. Neville, Manchester 1976.
17. Griffin, *Coalmining Industry*, 48; S. Pollard, *The Genesis of Modern Management: A Study of the Industrial Revolution in Great Britain*, Harmondsworth 1968, 80; Challinor and Ripley, *Miners' Association*, 195.
18. Clapham, *Economic History*, 121; Hair, 'Social History', 20; Challinor and Ripley, *Miners' Association*, 185; *Potts' Mining Register and Directory for the Coal and Ironstone Mines of Great Britain and Ireland. 1910*, North Shields n.d., 19—29; T. S. Ashton and J. Sykes, *The Coal Industry of the Eighteenth Century*, 2nd ed., Manchester 1964; *Reports of H.M. Inspectors of Mines*, 1864, 1884.
19. D. J. Rowe, 'The Economy of the North-East in the Nineteenth

Century: A Survey with a Bibliography of works published since 1945', *Northern History* VI (1971), 119; Pollard, *Genesis*, 80; N. R. Elliott, 'A Geographical Analysis of the Tyne Coal Trade', *Tijdschrift voor Economische en Sociale Geografie* 59/2 (1968), 73; R. Galloway, *A History of Coal Mining in Great Britain*, Newton Abbot 1969, 121–9.

20. *Colliery Guardian*, 9, 23 Feb. 1900; 2, 16 Mar. 1900; Hair, 'Social History', 17; Elliott, in *TESG* 59/2, 59, 73; P. J. Perry, *A Geography of 19th-Century Britain*, London 1975, 60.

21. Griffin, *Coalmining Industry*, 63; Rowe, in *Northern History* VI, 126; Galloway, *Coal Mining*, 203–210; Challinor and Ripley, *Miners' Association*, 146.

22. R. Moore, *Pit-Men*, 66.

23. *One Hundred Songs of Toil*, ed. K. Dallas, London 1974, 163.

24. Challinor, *Lancashire Miners*, 207, 260; Rowe, *Wages*, 19.

25. *Reports of H.M. Inspectors*, 1864; *Potts' Register 1910*, 52–62.

26. Challinor, *Lancashire Miners*, 49–50; Challinor and Ripley, *Miners' Association*, 185.

27. A. V. John, 'Women Workers in British Coal Mining, 1840–90, With Special Reference to West Lancashire', (Manchester Ph.D. thesis, 1976), 325, 327–8.

28. C. Storm-Clark, 'The Miners, 1870–1970: A Test-Case for Oral History', *Victorian Studies* XV/1 (Sept. 1971), 53.

29. Evison, 'Central Region', 5, 9, 12, 114, 125, 404; M. Jones, 'Changes in Industry Population and Settlement on the Exposed Coalfield of South Yorkshire 1840–1908', (Notthingham M.A. thesis, 1966), 55, 138; G. B. D. Gray, 'The South Yorkshire Coalfield', *Studies in the Yorkshire Coal Industry*, eds J. Benson and R. G. Neville, Manchester 1976, 35, 38; B. E. Coates and G. M. Lewis, *The Doncaster Area*, Sheffield 1966, 22.

30. Evison, 'Central Region', 118.

31. Evison, 'Central Region', 123–4; Jones, 'Exposed Coalfield', 53, 98.

32. Evison, 'Central Region', 10.

33. Gray, 'South Yorkshire', 36–7; Jones, 'Exposed Coalfield', 197.

34. Williams, *Derbyshire Miners*, 175–7; Griffin, *Coalmining Industry*, 164; Griffin, *East Midlands*, 97, 160; W. Smith, *An Economic Geography of Great Britain*, London 1949, 127; T. J. Chandler, 'Communications and a Coalfield: A Study in the Leicestershire and South Derbyshire Coalfield', *Transactions of the Institute of British Geographers* (1957), 163–5, 168–9; A. R. and C. P; Griffin, 'The Role of Coal Owners' Associations in the East Midlands in the Nineteenth Century', *Renaissance and Modern Studies* XVII (1973), 102; P. Renshaw, *The General Strike*, London 1975, 32; *Colliery Guardian*, 27 April, 1 June 1900.

35. Williams, *Derbyshire Miners*, 38, 177; *Potts' Mining Register and Directory for the Coal and Ironstone Trades of Great Britain and*

Ireland, 1889, North Shields 1889, 137—141, 173, 184, 202; *Potts' Register 1910,* 13—18, 63—4, 77—9, 97—8.

36. National Museum of Wales, *Welsh Coal Mines,* Cardiff 1976, 10; J. H. Morris and L. J. Williams, *The South Wales Coal Industry 1841— 1875,* Cardiff 1958, 1—10; John, *South Wales,* 1—2; H. S. Jevons, *The British Coal Trade,* London 1915, 101.

37. Morris and Williams, *South Wales,* 5, 12, 43—6, 91—8; Jevons, *Coal Trade,* 30, 94, 100—101; P. N. Jones, 'Aspects of the Population and Settlement Geography of the South Wales Coalfield 1850—1926', (Birmingham Ph.D. thesis, 1965), 38—40.

38. Morris and Williams, *South Wales,* 104. The growing domination of the export trade was strengthened by the stagnation of the iron industry after 1860 which further encouraged the ironmasters to produce coal for sale rather than for their own use.

39. Ashton and Sykes, *Coal Industry,* 9; Morris and Williams, *South Wales,* 134—7; John, *South Wales,* 141; R. Galloway, *Annals of Coal Mining and the Coal Trade* 2, Newton Abbot 1971, 349.

40. Morris and Williams, *South Wales,* 134—7.

41. Morris and Williams, *South Wales,* 152; Griffin, *Coalmining Industry,* 64—5; M. J. Daunton, *Coal Metropolis: Cardiff 1870—1914,* Leicester 1977, 60—61.

42. R. A. S. Redmayne, 'Coal Mining', *Victoria County History of Worcester* 2, eds J. W. Willis-Bund and W. Page, London 1971, 264; Hair, 'Social History', 16—17; A. J. Taylor, 'Coal', *Victoria County History of Staffordshire* 2, eds M. W. Greenslade and J. G. Jenkins, Oxford 1967, 71—7; R. W. Sturgess, 'Landownership, Mining and Urban Development in Nineteenth-Century Staffordshire', *Land and Industry: The Landed Estate and the Industrial Revolution,* eds J. T. Ward and R. G. Wilson, Newton Abbot 1971, 176—7.

43. Taylor, 'Coal', 77; E. Taylor, 'The Working Class Movement in the Black Country 1863—1914', (Keele Ph.D. thesis, 1974), 11, 209; G. C. Allen, *The Industrial Development of Birmingham and the Black Country 1860—1927,* London 1966, 237, 278—80.

44. Sturgess, 'Landownership', 177—8; Taylor, 'Coal' 70—71, 278—80.

45. Taylor, 'Coal', 71—9; Sturgess, 'Landownership', 178; Allen, *Birmingham,* 279—80.

46. G. Gilpin to J. Wise, 3 October 1819, *Iron in the Making: Dowlais Iron Company Letters 1782—1860,* ed. M. Elsas, Cardiff 1960, 7. Also Taylor, 'Coal', 69; Rowe, *Wages,* 147; Redmayne, 'Coal', 266; Pollard, *Genesis,* 85; Allen, *Birmingham,* 143.

47. Taylor, 'Coal', 98—9; Allen, *Birmingham,* 142; J. Benson, 'The Compensation of English Coal Miners and their Dependants for Industrial Accidents, 1860—97', (Leeds Ph.D. thesis, 1974), Table 3; G. J. Barnsby, 'Social Conditions in the Black Country in the Nineteenth Century', (Birmingham Ph.D. thesis, 1969), 31.

48. Taylor, 'Coal', 99; Benson, 'Compensation', Table 3; *Potts' Register, 1889,* 192—4.

49. R. A. Lewis, *Coal Mines in Staffordshire,* Stafford 1966, 27.

50. A. J. Youngson Brown, 'The Scots Coal Industry, 1854—1886'. (Aberdeen D.Litt. thesis, 1952), 38, 45, 66—8; Jevons, *Coal Trade*, 147, 150; W. Smith, *An Historical Introduction to the Economic Geography of Great Britain*, London 1968, 152—3; R. N. Millman, *The Making of the Scottish Landscape*, London 1975, 193.

51. J. A. Hassan, 'The Development of the Coal Industry in Mid and West Lothian 1815—1873', (Strathclyde Ph.D. thesis, 1976), 164; Youngson Brown, 'Coal Industry', 70, 95; *Potts' Register, 1889; Potts' Register, 1910.*

52. Youngson Brown, 'Coal Industry', 101—2; Hassan, 'Lothian', 134, 143; M. W. Kirby, *The British Coalmining Industry 1870—1946: A Political and Economic History*, London 1977, 8.

53. Youngson Brown, 'Coal Industry', 3; H. Hamilton, *The Industrial Revolution in Scotland*, London 1966, 169; J. H. G. Lebon, 'The Development of the Ayrshire Coalfield', *Scottish Geographical Magazine* 49 (1933), 140—2.

54. B. F. Duckham, *A History of the Scottish Coal Industry*, Newton Abbot 1970, I, 30; Lebon, 49, 143; Youngson Brown, 'Coal Industry', 6—7.

55. Jevons, *Coal Trade*, 146; *Colliery Guardian*, 1, 8, 22 June and 20 July 1900.

56. H. and B. Duckham, *Great Pit Disasters: Great Britain 1700 to the Present Day*, Newton Abbot 1973, 141; Youngson Brown, 'Coal Industry', 64, 72.

57. *Potts' Register, 1889*, 218—42; *Potts' Register, 1910*, 120—44.

58. *Potts' Register*, 120—22, 130—8; Youngson Brown, 'Coal Industry', 97—8.

59. R. E. Goffee, 'The Butty System and the Kent Coalfield', *Bulletin of the Society for the Study of Labour History* 34 (Spring 1977), 41.

60. Jevons, *Coal Trade*, 3.

61. Jevons, *Coal Trade*, 137; Smith, *Economic Geography*, 303.

62. C. G. Down and A. J. Warrington, *The History of the Somerset Coalfield*, Newton Abbot 1971, 17, 21; R. A. Buchanan and N. Cossons, *The Industrial Archaeology of the Bristol Region*, Newton Abbot 1969; Rowe, *Wages*, 29—31.

63. Griffin, *Coalmining Industry*, 30.

64. Professor Duckham's introduction to Galloway, *Coal Mining*.

65. Lewis, *Coal Mining*, 15.

Chapter 2
THE MINER AT WORK
(pp. 28—63)

1. Taylor, in *EHR* XIV/1, 68—9; Rowe, *Wages*, 15.

2. A. B. Campbell, 'Honourable Men and Degraded Slaves: A social history of trade unionism in the Lanarkshire coalfield, 1775—1874, with particular reference to the Coatbridge and Larkhall districts', (Warwick Ph.D. thesis, 1976), 47; Durham County Record Office,

D/NCB/CO/86(347), Durham Coal Owners' Association Return, 1894.

3. *Wolverhampton Chronicle*, 3 Nov. 1813.

4. J. Bullock, *Bowers Row: Recollections of a mining village*, Wakefield 1976, 47.

5. A. Potts and E. Wade, *Stand True*, Newcastle n.d., 4.

6. 'A Careful Observer' to *Miner & Workman's Advocate*, 17 Oct. 1863. Also 'A Man Knocking About Thornley', to *Miner & Workman's Advocate*, 3 Oct. 1863; *Report of the Select Committee on Acts for the Regulation and Inspection of Mines*, 1866, Q. 2,219, W. Pickard; B. McCormick and J. E. Williams, 'The Miners and the Eight-Hour Day, 1863—1913', *Economic History Review* XII/2 (1959), 237.

7. *Select Committee Mines*, Q. 654, P. Dickinson. Also Q. 1,790, W. Pickard; Q. 6, 846-50, A. M'Donald; J. Darton to *Miner & Workman's Advocate*, 28 Nov. 1863.

8. *Colliery Guardian*, 4 July 1873.

9. B. L. Coombes, *These Poor Hands: The Autobiography of a Miner Working in South Wales*, London 1939, 70.

10. *Colliery Guardian* 30 Jan. 1858.

11. John, 'Women', 426—7.

12. T. Dalkin, 'Story of 50 Yrs, A Miner, And Soldier'.

13. A. V. John, 'The Lancashire Pit Brow Lasses', *Bulletin of the North West Group for the Study of Labour History*, 3 (1976), 3; *Miner & Workman's Advocate*, 12 Dec. 1863; John, 'Women', 430.

14. Duckham, *Disasters*, 15.

15. *Select Committee Mines*, Q. 1,728, W. Pickard; Q. 6, 835—8, A. M'Donald; *Colliery Guardian*, 5 May 1905; Campbell, 'Slaves', 252; John, 'Women', 263; *Report of the Commissioner Appointed Under the Provisions of the Act 5 & 6 Vict. c.99, . . .*, 1844, 50—51; *Reports of H.M. Inspectors of Mines*, R. McLaren, 1904.

16. *Select Committee Mines*, Q. 9,813, Rev. H. R. Sandeford; J. Wilson, *Autobiography*, 19; G. Parkinson, *True Stories of Durham Pit-Life*, London 1912, 16, 19; J. Halliday, *Just Ordinary, But . . . an Autobiography*, Waltham Abbey 1959, 63.

17. W. Bell, 'Reminiscences', 45.

18. Bullock, *Bowers Row*, 215.

19. 'The autobiography of Edmund Stonelake', ch. 7, 6; 'W.E.' to *British Miner and General Newsman*, 13 Dec. 1862; Halliday, *Just Ordinary*, 68—9; Griffin, *Coalmining Industry*, 120—1; Duckham, *Disasters*, 21—3; Duckham's introduction to Galloway, *Coal Mining;* Galloway, *Annals*, I, 488; Jevons, *Coal Trade*, 398—400.

20. Parkinson, *True Stories*, 23; Bullock, *Bowers Row*, 184.

21. Duckham's introduction to Galloway, *Coal Mining;* T. L. Llewellyn, *Miners' Nystagmus: Its Causes and Prevention*, London 1912, 133; Evison, 'Central Region', 140.

22. W. Bell, 'Reminiscences', 43.

23. Halliday, *Just Ordinary*, 69. Also W. Bell, 'Reminiscences', 46;

Colliery Guardian, 25 Feb. 1910; Northumberland Miners' Mutual Confident Association, *Minutes, Monthly Circular*, Nov. 1903; Hair, 'Social History', 161.

24. D. Douglass, *Pit Talk in County Durham: A Glossary of Miners' Talk Together With Memories of Wardley Colliery, Pit Songs and Piliking*, Oxford 1973, 22–3; also Jevons, *Coal Trade*, 615; Youngson Brown, 'Coal Industry', 249.

25. Galloway, *Annals* 2, 354; A. Burton, *The Miners*, London 1976, 76; Llewellyn, *Nystagmus*, 51; *Colliery Guardian*, 20 Sep. 1872.

26. A. J. Parfitt, *My Life as a Somerset Miner*, Radstock 1930, 12.

27. Kirby, *Coalmining*, 7.

28. Llewellyn, *Nystagmus*, 47.

29. 'A Somerset' to *Miner & Workman's Advocate*, 28 Nov. 1863; *Select Committee Mines*, Q. 219, T. Burt; *Colliery Guardian* 9, 16 Feb. 1900; Hair, 'Social History', 125; Youngson Brown, 'Coal Industry', 249; Griffin, *Coalmining Industry*, 51; Jevons, *Coal Trade*, 75; Galloway, *Annals* 2, 348, 350.

30. Challinor and Ripley, *Miners' Association*, 49; Galloway, *Annals* 2, 353–4.

31. *Wolverhampton Chronicle*, 16 Nov. 1825.

32. E. A. Rymer, 'The Martyrdom of the Mine, or, A 60 Years' Struggle for Life', ed. R. G. Neville, *History Workshop* 1 (Spring 1976), 4.

33. W. Bell, 'Reminiscences', 45.

34. Challinor and Ripley, *Miners' Association*, 148–9.

35. J. Tunstall, *The Fishermen*, London 1962, 22–3.

36. *Barnsley Chronicle*, 3 Mar. 1877; *Colliery Guardian*, 5 July 1878, 2 Feb. 1900.

37. *Colliery Guardian*, 20 Apr. 1900.

38. But see the important article by P. E. H. Hair, 'Mortality from Violence in British Coal-Mines, 1800–50', *Economic History Review* XXI/3 (1968).

39. Morris and Williams, *South Wales*, 195. See also Hair, 'Social History', 92, 108; Evison, 'Central Region', 26; *Colliery Guardian*, 19 May 1905.

40. See Statistical Appendix.

41. Duckham, *Disasters*, 29.

42. Duckham, *Disasters*, 16.

43. Parkinson, *True Stories*, 42.

44. G. L. Campbell, *Miners' Insurance Funds: Their Origin and Extent*, London 1880, 6.

45. *Barnsley Chronicle*, 6 Oct., 27 Jan. 1877; *Barnsley Independent*, 2 Nov. 1878

46. *Children's Employment Commission, 1st Report*, 1842, 144.

47. Evison, 'Central Region', J. Evison, 'Conditions of Labour in Yorkshire Coal-Mines, 1870–1914', (Birmingham B.A. thesis, 1963), 12.

48. The following section is based on my 'Non-fatal coalmining accidents', *Bulletin of the Society for the Study of Labour History*, 32

(Spring 1976). A non-fatal accident is defined as an injury resulting in at least one day's absence from work.

49. *Reports of H.M. Inspectors of Mines*, J. J. Atkinson, 1864, 30. See also J. Gerrard, 1896, 10–11; 1897, 10; A. H. Stokes, 1897, 17.

50. *Reports of H.M. Inspectors of Mines*, 1896 and 1897; B. R. Mitchell and P. Deane, *Abstract of British Historical Statistics*, Cambridge 1962, 72; *The Colliery Year Book and Coal Trades Directory 1938*, 634.

51. Stonelake, 'autobiography', ch. 5, 3.

52. *Reports of H.M. Inspectors of Mines*, 1889–90.

53. Jevons, *Coal Trade*, 122–3; R. A. Redmayne, *Men, Mines and Memories*, London 1942, 166–7; Rowe, *Wages*, 21; Hair, in *EHR* XXI/3, 549; Griffin, *Coalmining Industry*, 101.

54. Hair, in *EHR* XXI/3, 549.

55. *Colliery Guardian*, 17 July 1858; Challinor *Lancashire Miners*, 158–9.

56. Hair, in *EHR* XXI/3, 560; Central Association for Dealing with Distress Caused by Mining Accidents, *1891 Report*, 54.

57. R. Ridyard, *Mining Days in Abram*, Leigh 1972, 37.

58. 'Veritas' to *British Miner And General Newsman*, 11 Oct. 1862.

59. Dallas, *Songs*, 210.

60. J. L. Cohen, *Workmen's Compensation in Great Britain*, London 1923, 169–70.

61. *Reports of H.M. Inspectors of Mines*, H. Mackworth, 1856, 117; Northumberland Miners' Mutual Confident Association, *Minutes, Circular*, Mar. 1904; *Select Committee Mines*, Q. 572, W. Crawford; Q. 3,532, J. Normansell; Durham County Record Office, D/MRP 60/5, Mrs A. Hodges, 'I remember', 66; Hair, 'Social History', 109.

62. D. Douglass, *Pit Talk in County Durham; A Glossary . . .*, Oxford 1973, 36.

63. See for example *Colliery Guardian*, 24 Mar. 1905.

64. *Miner & Workman's Advocate*, 19 Sep. 1863.

65. *Miner & Workman's Advocate*, 19 Sep. 1863; Hair, 'Social History', 94, 103, 105–6.

66. Hair, 'Social History', 104–5; *Colliery Guardian*, 12 Sep. 1863.

67. A. Meiklejohn, 'History of Lung Diseases of Coal Miners in Great Britain: Part II, 1875–1920', *British Journal of Industrial Medicine*, 9/2 (1952), 94–5.

68. A. Horner, *Incorrigible Rebel*, London 1960, 11.

69. A. Wilson and H. Levy, *Workmen's Compensation 2*, Oxford 1939, 87.

70. Hair, 'Social History', 104–6; Down and Warrington, *Somerset*, 36; *Colliery Guardian*, 12 Sep. 1863; Northumberland Miners' Mutual Confident Association, *Minutes, Monthly Circular*, Oct. 1903.

71. Llewellyn, *Nystagmus*, 1, 2, 134; Wilson and Levy, *Compensation*, 2, 220.

72. Williams, *Derbyshire Miners*, 478–80; Llewellyn, *Nystagmus*, 26–36.

73. Llewellyn, *Nystagmus*, 15, 131—5.
74. Llewellyn, *Nystagmus*, 6—7.
75. *British Miner And General Newsman*, 18 Oct. 1862. Also *Colliery Guardian*, 18 Feb. 1910.
76. R. G. Neville, 'The Yorkshire Miners, 1881—1926: A Study in Labour and Social History', (Leeds Ph.D. thesis, 1974), 818—9.
77. *Reports of H.M. Inspectors of Mines*, W. N. Atkinson, 1903, 29; 1904, 27.
78. G. M. Wilson, 'The Miners of the West of Scotland and their Trade Unions, 1842—74', (Glasgow Ph.D. thesis, 1977), 42.
79. Interview with A. J. R. Hickling, Sedgley, 18 Dec. 1977.
80. Horner, *Rebel*, 11.
81. Taylor, in *EHR* XIV/1, 57—8; Evison, 'Central Region', 146; Campbell, 'Slaves', 41.
82. *Select Committee Mines*, Q. 2, 873; 2, 887, R. Woodward.
83. Rymer, in *HW* 1, 4.
84. *Select Committee Mines*, Q. 3,029, J. Normansell.
85. *Select Committee Mines*, Q. 1,177, W. Baxendale.
86. By the end of the century trappers were almost redundant because brattice cloths were introduced which the ponies were able to push aside.
87. G. Mee, *Aristocratic Enterprise: The Fitzwilliam Industrial Undertakings 1795—1857*, Glasgow 1975, 132—3; Taylor, 'Coal', 102: *Report of Commissioner*, 1853, 43—4; *Select Committee Mines*, Q. 8, T. Burt; Hair, 'Social History', 166, 173—4, 212.
88. Parkinson, *True Stories*, 10—11.
89. *Select Committee Mines*, Q. 143, T. Burt; Rymer, in *HW* 1, 4.
90. Parkinson, *True Stories*, 23.
91. Morris and Williams, *South Wales*, 211.
92. Rymer, in *HW* 1, 8.
93. Storm-Clark, in *VS* XV/1, 56.
94. Douglass, *Pit Talk*, 76.
95. Morris and Williams, *South Wales*, 212.
96. Douglass, *Pit Talk*, 220.
97. Hair, 'Social History', 252; Challinor, *Lancashire Miners*, 253; Morris and Williams, *South Wales*, 215. But for the continued employment of women underground see for example *Colliery Guardian*, 6 Mar. 1858.
98. Morris and Williams, *South Wales*, 214.
99. Duckham, *Scottish Industry*, 96.
100. Hair, 'Social History', 205; Taylor, in *EHR* XIV/1, 57—8; *Report of Commissioner*, 1853, 20; Griffin, *East Midlands*, 14; Rymer, in *HW* 1, 5.
101. D. Douglass, 'The Durham pitman', *Miners, Quarrymen and Salt-workers*, ed. R. Samuel, London 1977, 220; Hair, 'Social History', 53, 209; *Select Committee Mines*, Q. 3,098, J. Normansell; Q. 6,803, A. M'Donald.
102. K. Durland, *Among the Fife Miners*, London 1904, 28—30.

103. Rymer, in *HW* 1, 4.

104. Hair, 'Social History', 197—8; Morris and Williams, *South Wales*, 212; Neville, 'Yorkshire Mines', 277, 282.

105. Wilson, *Autobiography*, 20. Also Hair, 'Social History', 197—8; Morris and Williams, *South Wales*, 212.

106. *British Miner And General Newsman*, 13 Dec. 1862; N. Milburn to *The Miner*, 23 May 1863.

107. *Miners' Advocate*, 26 July 1873; Wilson, *Autobiography*, 21; *Children's Employment Commission, Appendix to 1st Report*, Part 1, 255.

108. Parkinson, *True Stories*, 1.

109. Hair, 'Social History', 164, 212, 384; Challinor, *Lancashire Miners*, 12; M. Benney, *Charity Main: A Coalfield Chronicle*, London 1946, 100, 171.

110. Douglass, *Pit Talk*, 66—7.

111. Parkinson, *True Stories*, 1. Also Hair, 'Social History', 138—9, 211—12; Morris and Williams, *South Wales*, 236.

112. *Colliery Guardian*, 1 June 1900; Hair, 'Social History', 123, 140; Youngson Brown, 'Coal Industry', 58; R. Walters, 'Labour Productivity in the South Wales Steam-Coal Industry, 1870—1914', *Economic History Review* XXVIII/2 (May 1975), 283—4; Campbell, 'Slaves', 38, 41, 44, 62—3, 160; Evison, 'Central Region', 145.

113. Stonelake, 'autobiography', ch. 9, 2.

114. Burgess, *Industrial Relations*, 101; Griffin, *Coalmining Industry*, 7, 10; Hair, 'Social History', 135—6; Youngson Brown, 'Coal Industry', 71.

115. Hair, 'Social History', 136.

116. Jevons, *Coal Trade*, 209—10.

117. R. S. Sayers, *A History of Economic Change in England, 1880—1939*, London 1967, 87.

118. Walters, in *EHR* XXVIII/2, 296—8; Duckham's introduction to Galloway, *Coal Mining*; Smith, *Historical Introduction*, 160; Taylor, in *EHR* XIV/1, 60.

119. Parkinson, *True Stories*, 1.

120. *Select Committee Mines*, Q. 163, T. Burt; Dallas, *Songs*, 194.

121. Morris and Williams, *South Wales*, 277—9.

122. McCormick and Williams, in *EHR* XII/2, 236. See also Hair, 'Social History' 149; Youngson Brown, 'Coal Industry', 56; Barnsby, 'Social Conditions', 57—8, 307; *Colliery Guardian*, 20, 27 June 1879.

123. *Miners' Advocate*, 5 Apr. 1873; *Select Committee Mines*, Q. 3,151, J. Normansell.

124. *Colliery Guardian*, 22 Aug. 1879; Hair, 'Social History', 178—9, 426; Duckham, *Scottish Industry*, 73, 311; Campbell, 'Slaves', 73; Hassan, 'Lothian', 242.

125. Bullock, *Bowers Row*, 204; Walters, in *EHR* XXVIII/2, 290.

126. Hassan, 'Lothian', 328. Also Hair, 'Social History', 149 n.2.

127. *Colliery Guardian*, 17 Aug. 1900.

128. *Colliery Guardian*, 25 Oct. 1872. Also 5 July 1872; Neville, 'Yorkshire Miners', 273; Campbell, 'Slaves', 233.

129. R. Colls, *The Collier's Rant: Song and Culture in the Industrial Village*, London 1977, 201, n.8.

130. *Colliery Guardian*, 5 July 1878.

131. *Colliery Guardian*, 5 Sep. 1879. See also 15 Aug., 12 Sep. 1873; 27 June, 1, 29 Aug., 5 Sep., 10 Oct. 1879; 6 Jan. 1905; 28 Jan. 1910.

132. *Colliery Guardian*, 14, 21 Feb. 1879.

133. *Colliery Guardian*, 6 Jan. 1905; University College, Swansea, South Wales Miners' Federation, *Minutes*, Council Meeting, 1910.

134. *Report of Commissioner*, 1853, 23. Also *Select Committee Mines*, Q. 414, T. Burt; *Colliery Guardian*, 23 Aug. 1872.

135. *Wolverhampton Chronicle*, 14 Feb. 1900. Also D. A. Reid, 'The Decline of Saint Monday 1766—1876', *Past and Present* 71 (Aug. 1976), 91.

136. *Colliery Guardian*, 10 Jan. 1879.

137. *Colliery Guardian*, 21 May 1880. See also 14 Jun. 1878; 18 Apr., 6 June 1879; 2 Apr. 1880.

138. Taylor, in *EHR* XIV/1, 54; Walters, in *EHR* XXVII/2, 293.

139. Taylor, in *EHR* XIV/1, 53.

140. Williams, *Derbyshire Miners*, 61. Compare N. Dennis, F. Henriques and C. Slaughter, *Coal is our Life: An Analysis of a Yorkshire Mining Community*, London 1956, 186.

141. I have been unable to discover any firm evidence to support this idea, but it does seem one that may be worth pursuing.

142. See for example F. Machin, *The Yorkshire Miners: A History* 1, Barnsley 1958 and R. Page Arnot, *The Miners: A History of the Miners' Federation of Great Britain 1889—1910*, London 1949.

143. *Colliery Guardian*, 20 June 1879; 24 Aug. 1900; 3 Feb. 1905; Morris and Williams, *South Wales*, 192; Parfitt, *Somerset Miner*, 19.

144. Walters, in *EHR* XXVIII/2, 293.

145. Campbell, 'Slaves', 139.

146. *Colliery Guardian*, 13 Mar. 1858.

147. *Wolverhampton Chronicle*, 9 Aug. 1837.

148. *Colliery Guardian*, 11 Oct. 1872; 4 July 1873.

149. Northumberland Miners' Mutual Confident Association, *Minutes*, Committee Meeting, 30 Nov. 1881; *Colliery Guardian*, 1 Mar. 1878.

150. *Colliery Guardian*, 12 June 1858; Evison, 'Central Region', 51, Griffin, *Coalmining Industry*, 159.

151. Morris and Williams, *South Wales*, 232—3.

152. 'W.E.' to *Miner And Workman's Advocate*, 13 June 1863.

153. *Colliery Guardian*, 4 June 1880.

154. Northumberland Miners' Mutual Confident Association, *Minutes*, Committee Meeting, 19 Feb. 1881.

155. *Colliery Guardian*, 3 Feb. 1905.

156. Walters, in *EHR* XXVIII/2, 293; Taylor in *EHR* XIV/1, 70.

157. Interview with H. J. Evans, Essington, 29 Dec. 1977; J. Sheldon

to *British Miner And General Newsman,* 15 Nov. 1862.

158. *Select Committee Mines,* Q. 2, 852—3, R. Woodward.

159. Dallas, *Songs,* 218.

Chapter 3
THE MINER'S EARNINGS
(pp. 64—80)

1. B. Holman, *Behind the Diamond Panes: The Story of a Fife Mining Community,* Cowdenbeath 1952, 87.

2. Griffin, *Coalmining Industry,* 78; Campbell, 'Slaves', 26—35.

3. Wilson, *Autobiography,* 88; South Wales Miners' Federation, *Minutes,* Council Meeting, 30 Mar. 1908.

4. Hodges, 'I remember', 4; Homan, *Fife,* 49.

5. Griffin, *Coalmining Industry,* 84—5; Challinor, *Lancashire Miners,* 27; Ashton and Sykes, *Coal Industry,* 142; Staffordshire County Record Office, D. 593/M/2/2/2, J. Loch, 26 Oct. 1818.

6. Coombes, *Poor Hands,* 45.

7. R. Mitchell to *Miner And Workman's Advocate,* 4 July 1863. Also W. J. Bell to *Miner And Workman's Advocate,* 19 Dec. 1863; Northumberland Miners' Mutual Confident Association, *Minutes,* Committee Meeting, 15 Dec. 1885; *Colliery Guardian,* 10 Apr. 1879; 9 Apr. 1880; 2 June 1905; 'An Old Collier' to *The Miner,* 14 Mar. 1863; *Select Committee Mines,* Q. 3, 218, J. Normansell.

8. Campbell, 'Slaves', 153.

9. *Colliery Guardian,* 15 Mar. 1876; Rowe, *Wages,* 28, 63.

10. Hair, 'Social History', graph 1.

11. Youngson Brown, 'Coal Industry', 228; Hassan, 'Lothian', 244.

12. *Colliery Guardian,* 4 Apr. 1879.

13. *Colliery Guardian,* 21 June 1878; 4 June 1880; Barnsby, 'Social Conditions', 301; Williams, *Derbyshire Miners,* 58; Rymer, in *HW* 2, 20; *Wakefield Express,* 28 Dec. 1878.

14. Challinor, *Lancashire Miners,* 211; John, in *BNWGSLH* 3, 4.

15. Griffin, *Coalmining Industry,* 78.

16. *Colliery Guardian,* 1 Nov. 1872; 18 Jun. 1880; *Miners' Advocate,* 7 Mar. 1874; Potts and Wade, *Stand True,* 8.

17. John, 'Women', 188 n.1.

18. Hassan, 'Lothian', 234—5. See also Campbell, 'Slaves', 62.

19. Hair, 'Social History', 53, 209.

20. Storm-Clark, in *VS* XV/1, 56.

21. Rymer, in *HW* 1, 5.

22. G. C. Greenwell, *A Glossary of Terms Used in the Coal Trade of Northumberland and Durham,* London 1888, 6.

23. Dalkin, 'Story'.

24. B. Thomas, 'The Migration of Labour into the Glamorganshire Coalfield, 1861—1911', *Industrial South Wales 1750—1914: Essays in Welsh Economic History,* ed. W. E. Minchinton, London 1969, 42.

25. Rowe, *Wages,* 82. See also Hassan, 'Lothian', 235; Evison, 'Central

Region', 263; Challinor, *Lancashire Miners*, 211—12; Williams, *Derbyshire Miners*, 439.

26. Taylor, 'Black Country', 121; Youngson Brown, 'Coal Industry', 220.
27. Jevons, *Coal Trade*, 338—9.
28. Hair, 'Social History', 383; J. R. Raynes, *Coal and its Conflicts: A Brief Record of the Disputes between Capital & Labour in the Coal Mining Industry of Great Britain*, London 1928, 29.
29. Wilson, *Autobiography*, 24—5.
30. *Colliery Guardian*, 30 May 1879. See also 2 Apr. 1880; 12 Apr. 1900; 14 Apr. 1905; Hodges, 'I remember', 67; Challinor and Ripley, *Miners' Association*, 58 n.1.
31. *Colliery Guardian*, 8 Aug. 1873.
32. Rowe, *Wages*, 82.
33. Burgess, *Industrial Relations*, 163.
34. *Colliery Guardian*, 10 Mar. 1905.
35. Douglass, *Pit Talk*, 239.
36. Parfitt, *Somerset Miner*, 18.
37. 'Tainwell' to *Miner & Workman's Advocate*, 14 Nov. 1863.
38. Neville, 'Yorkshire Miners', 297.
39. Parfitt, *Somerset Miner*, 18.
40. Griffin, *Coalmining Industry*, 33; Rowe, *Wages*, 151; Taylor, 'Black Country', 209—11.
41. Staffordshire County Record Office, D 593 K/1/8/6, Florence Colliery Memorandum, c. 1900; Taylor, 'Black Country', 123.
42. John, 'Women', 271—5; *Select Committee Mines*, Q. 2,323—4, W. Pickard.
43. Rowe, *Wages*, 84; *Colliery Guardian*, 30 June 1905; Northumberland Miners' Mutual Confident Association, *House-Rent and Fire-Coal Question*, 1881; *Minutes*, Council Meeting, 21 Nov. 1903.
44. *Colliery Guardian*, 16 Jan. 1858; 25 Apr. 1879. See too Rymer, in *HW* 1, 6; 'J.F.'. to *Miner & Workman's Advocate*, 26 Dec. 1863; Evans interview.
45. Rowe, *Wages*, 84.
46. Campbell, 'Slaves', 101; *Select Committee Mines*, Q. 9, 320, T. Evans.
47. Evison, 'Cental Region', 263, 265.
48. Taylor, 'Coal', 103; Jevons, *Coal Trade*, 355—6.
49. Rowe, *Wages*, 51.
50. Hair, 'Social History', 322; Campbell, 'Slaves', 148; *Miner & Workman's Advocate*, 15 Aug. 1863; 'Ignoramus' to *Miner & Workman's Advocate*, 25 July 1863; G. W. Hilton, 'The Truck Act of 1831', *Economic History Review* X/3 (1958), 470; Wilson, 'Miners', 69.
51. 'Scotas' to *British Miner And General Newsman*, 31 Jan. 1863.
52. *Select Committee Mines*, Q. 7,131—40, A. McDonald. Also Youngson Brown, 'Coal Industry', 233; A. M'Donald to *Miner & Workman's Advocate*, 20 June 1863.
53. 'Glowr' to *British Miner And General Newsman*, 31 Jan. 1863.

54. R. Challinor, *Alexander MacDonald and the Miners*, London 1968, 9.

55. Stonelake, 'autobiography', ch. 2, 13.

56. 'An Observer' to *Wolverhampton Chronicle*, 9 Oct. 1816.

57. B. Penn to *Wolverhampton Chronicle*, 24 Apr. 1822.

58. Lewis, *Coal Mining*, 32.

59. Morris and Williams, *South Wales*, 231; Hassan, 'Lothian', 229—31; Lewis, *Coal Mining*, 32—33.

60. The following section is based on Rowe, *Wages*, 82—5; Hair, 'Social History', 316, 337—67; Jevons, *Coal Trade*, 58—154; Challinor, *Lancashire Miners*, 287 n.45; *Colliery Guardian*, 25 Apr. 1879; Duckham, *Scottish Industry*, 310—11.

61. Benson, 'Compensation', 218—28.

62. Hair, 'Social History', 392; Ashton and Sykes, *Coal Industry*, 94, 110; Griffin, *Coalmining Industry*, 70.

63. Taylor, 'Coal', 102.

64. Griffin, *East Midlands*, 113. See also Hair, 'Social History', 372—4; Morris and Williams, *South Wales*, 214.

65. *Colliery Guardian*, 4 June 1880.

66. Griffin, *Coalmining Industry*, 78.

67. John, 'Women', 96.

68. A. L. Lloyd, *Folk Song in England*, St Albans 1975, 329.

69. Rowe, *Wages*, 85, 152—5; Hair, 'Social History', 338—44; Taylor, in *EHR* XIV/1, 68—9.

70. Hair, 'Social History', 358—60; Morris and Williams, *South Wales*, 223; Ashton and Sykes, *Coal Industry*, 138—9.

71. Ashton and Sykes, *Coal Industry*, 139; Rowe, *Wages*, 85.

72. The following discussion is based on Griffin, *Coalmining Industry*, 76—80; Gregory, *Politics*, 180; P. Mathias, *The First Industrial Nation: An Economic History of Britain 1700—1914*, London 1969, 217; R. Page Arnot, *South Wales Miners Glowyr de Cymru: A History of the South Wales Miners' Federation (1898—1914)*, London 1967, 276; Evison, 'Central Region', 425.

Chapter 4
THE MINING SETTLEMENT
(pp. 81—111)

1. This section is based upon N. McCord and D. J. Rowe, 'Industrialisation and Urban Growth in North-East England', *International Review of Social History* XXII/1 (1977); Hetton Town Council, *Hetton Town: Official Guide*, Hetton-le-Hole n.d.; J. Y. E. Seeley, 'Coal Mining Villages of Northumberland and Durham: A Study of Sanitary Conditions and Social Facilities, 1870—1880', (Newcastle M.A. thesis, 1973); Sturgess, *Aristocrat*; J. H. Griffiths and F. Rundle, *The Pit on the Downs: A History of Eppleton Colliery 1825—1975*, Houghton-le-Spring n.d.

2. Redmayne, *Memories*, 8.

238 British Coalminers in the Nineteenth Century

3. Lewis, *Coal Mining*, 29. See too Burton, *Miners;* Neville, 'Yorkshire Miners', 806.
4. Griffin, *Coalmining Industry*, 132; Neville, 'Yorkshire Miners', 806.
5. *Select Committee Mines*, Q. 7, 147, A. M'Donald.
6. Griffin, *Coalmining Industry*, 157.
7. Hair, 'Social History', 21; Storm-Clark, in *VS* XV/1, 64; Jones, 'Population', 225.
8. McCord and Rowe, in *IRSH* XXII/1, 33.
9. *Colliery Guardian*, 27 Apr. 1900.
10. Gregory, *Politics*, 63. For South Wales see Morris and Williams, *South Wales*, 240 and for North Staffordshire see *Report of Commissioner*, 1850, 7.
11. M.I.A. Bulmer, 'Sociological Models of the Mining Community', *Sociological Review* 23/1 (Feb. 1975), 61.
12. Hair, 'Social History', 23–4, 83 n.1.; Down and Warrington, *Somerset*, 37.
13. Jones, 'Exposed Coalfield', 101.
14. Jevons, *Coal Trade*, 621.
15. *Colliery Guardian*, 13 June 1879. Also Griffin, *Coalmining Industry*, 163.
16. Jevons, *Coal Trade*, 655–6.
17. J. MacFarlane, 'Denaby Main: A South Yorkshire Mining Village', *Studies in the Yorkshire Coal Industry*, eds J. Benson and R. G. Neville, Manchester 1976, 112–3.
18. Duckham, *Disasters*, 141.
19. Gray, 'South Yorkshire', 38.
20. *Potts' Register*, 1889, 207–14.
21. Halliday, *Just Ordinary*, 88.
22. Gregory *Politics*, 2. Also Jones, 'Population', 6–7.
23. Coates and Lewis, *Doncaster*, 34–5; P. N. Jones, *Colliery Settlement in the South Wales Coalfield 1850 to 1926*, Hull 1969, 98; Jones, 'Exposed Coalfield', 89; J. Davison, *Northumberland Miners 1919–1939*, Newcastle 1973, 205.
24. Halliday, *Just Ordinary*, 15.
25. Jevons, *Coal Trade*, 651.
26. Seeley, 'Villages', 45, 57.
27. Seeley, 'Villages', 288.
28. Halliday, *Just Ordinary*, 16.
29. *Report of Commissioner*, 1853, 25.
30. Griffin, *Coalmining Industry*, 166.
31. Jevons, *Coal Trade*, 655–6.
32. Campbell, 'Slaves', 56.
33. Storm-Clark, in *VS* XV/1, 64–5.
34. For help with this section I am grateful to Mr G. E. Milburn of Sunderland Polytechnic.
35. Moore, *Pit-Men*, 82; H. Goodley, 'Bewicke Main. The Colliery which Disappeared', 30.

36. *Report of Commissioner*, 1850, 68.
37. *Report of Commissioner*, 1844, 48.
38. Goodley, 'Bewicke Main', 30. Also Bullock, *Bowers Row*, 119.
39. Hair, 'Social History', 414—5; Seeley, 'Villages', 48.
40. *Report of Commissioner*, 1850, 75. Also Seeley, 'Villages', 85.
41. Wilson, *Autobiography*, 89—90.
42. Coombes, *Poor Hands*, 21, 142. Also Holman, *Fife*, 90—91.
43. Evans interview.
44. Durham County Record Office, D/MRP 5/5(i), G. McBurnie, 'Washington 1910—1974'.
45. Hodges, 'I remember', 94. Also Holman, *Fife*, 88—9; Bullock, *Bowers Row*, 224.
46. Seeley, 'Villages', 127.
47. *Colliery Guardian*, 30 Jan. 1858.
48. Evison, 'Central Region', 421.
49. Seeley, 'Villages', 152. See also 'A Working Man' to *British Miner And General Newsman*, 18 Oct. 1862.
50. *Select Committee Mines*, Q. 7,129, A. M'Donald.
51. Seeley, 'Villages', 274.
52. Seeley, 'Villages', 219.
53. 'Veritas' to *British Miner And General Newsman*, 18 Oct. 1862.
54. *Miner & Workman's Advocate*, 18 July 1863.
55. Evans interview. See also *Hamilton Advertiser*, 15 Dec. 1877; 29 Dec. 1879; 26 Feb. 1887.
56. P. H. J. H. Gosden, *Self-Help: Voluntary Associations in Nineteenth-Century Britain*, London 1973, 185, 195—6.
57. *Miner & Workman's Advocate*, 18 July 1863; Evison, 'Central Region', 426.
58. T. Messer to *British Miner And General Newsman*, 25 Oct. 1862.
59. 'Live And Let Live' to *British Miner And General Newsman*, 10 Jan. 1863; 'Justice' to *The Miner & Workman's Advocate*, 26 Sept. 1863; Seeley, 'Villages', 125.
60. Seeley, 'Villages', 61; *Hetton Handbook*, 12; Durland, *Fife*, 137; information received from the Northumberland Room of Newcastle Central Library.
61. *Colliery Guardian*, 6 Apr. 1900.
62. 'T—' to *Miner & Workman's Advocate*, 17 Jan. 1863.
63. J. Lawson, *Who Goes Home?*, London 1945, 85. Also Durland, *Fife*, 137.
64. Jevons, *Coal Trade*, 631.
65. Hassan, 'Lothian', 298, 302; *Colliery Guardian*, 15 Nov. 1872; 23 Feb. 1900; Northumberland Miners' Mutual Confident Association, *Minutes*, Committee Meeting, 26 May 1881; Jevons, *Coal Trade*, 645.
66. *Colliery Guardian*, 25 Apr. 1879; 9 Mar. 1900; Evison, 'Central Region', 280.
67. Morris and Williams, *South Wales*, 240—1; Jevons, *Coal Trade*, 646; Jones, 'Population', 173, 204, 275, 294.

68. Griffin, *Coalmining Industry*, 161–2.
69. Mee, *Enterprise*, 141.
70. Hassan, 'Lothian', 298, 302; see too *Report of Commissioner*, 1855, 19; Youngson Brown, 'Coal Industry', 255.
71. *Report of Commissioner*, 1853, 25.
72. Bulman, *Coal Mining*, 274.
73. Hassan, 'Lothian', 302.
74. Staffordshire County Record office, D 593/M/2/2/2, Order signed J. Lock, 26 Oct. 1818. Also *Report of Commissioner*, 1846, 43.
75. *Report of Commissioner*, 1853, 30; Moore, *Pit-Men*, 84; Holman, *Fife*, 43.
76. *Report of Commissioner*, 1853, 25; 1859, 56.
77. *Report of Commissioner*, 1853, 25, 52.
78. Williams, *Derbyshire Miners*, 443.
79. *Report of Commissioner*, 1844, 10; 1859, 8; Jevons, *Coal Trade*, 648; J. C. M'Vail, *Housing of Scottish Miners: Report on the Housing of Miners in Stirlingshire and Dunbartonshire*, Glasgow 1911, 23; Holman, *Fife*, 22.
80. *Report of Commissioner*, 1859, 58.
81. Jevons, *Coal Trade*, 650–2; Youngson Brown, 'Coal Industry', 251.
82. *Miner & Workman's Advocate*, 19 Sept. 1863.
83. Wilson, *Autobiography*, 46.
84. Morris and Williams, *South Wales*, 239; Evans interview.
85. Staffordshire County Record Office, D 593 N/3/2/16.
86. Seeley, 'Villages', 292.
87. Northumberland County Record Office, NRO 1284/NMRS 4, Transcript of interview with Mr Billy Brown, 6 July 1976.
88. Halliday, *Just Ordinary*, 15–16.
89. Durham County Record Office, *Report of the Whickham Urban District Council as to the Condition of Marley Hill and District . . .*, June 1896, 16.
90. Seeley, 'Villages', 21.
91. Jevons, *Coal Trade*, 63; Seeley 'Villages', 293: *Miners' Advocate*, 21 Mar. 1874.
92. Rymer in *HW* 1, 8.
93. *Miners' Advocate*, 9 Aug. 1873.
94. Seeley, 'Villages', 54.
95. M'Vail, *Housing*, 46.
96. *Report of Commissioner*, 1850, 14.
97. M'Vail, *Housing*, 45.
98. Rymer, in *HW* 2, 24.
99. Evison, 'Central Region', 321, 328.
100. Evans interview; also Lewis, *Coal Mining*, 38; Seeley, 'Villages', 94, 131.
101. *Wolverhampton Chronicle*, 24 July 1850.
102. Seeley, 'Villages', 312–3.
103. Seeley, 'Villages', 59.

104. Brown interview; Goodley, 'Bewicke Main', 10.
105. Evison, 'Central Region', 337.
106. C. Evans, *The Industrial and Social History of Seven Sisters*, Cardiff 1964, 56.
107. Bullock, *Bowers Row*, 10–11.
108. Hassan, 'Lothian', 299.
109. Evison, 'Central Region', 296.
110. *Newcastle Weekly Chronicle*, 26 Oct. 1872.
111. Seeley, 'Villages', 47.
112. M'Vail, *Housing*, 39.
113. T. Thomas to *British Miner And General Newsman*, 28 Feb. 1863.
114. Davison, *Northumberland Miners*, 208–9; Seeley, 'Villages', 4, 302–3.
115. Halliday, *Just Ordinary*, 16.
116. Durham County Record Office, D/MRP 161/2, Interview with Rose Larmer.
117. Seeley, 'Villages', 303.
118. Seeley, 'Villages', 112.
119. Durham County Record Office, D/MRP 158/1, 22.
120. Hodges, 'I remember', 108. Also Bullock, *Bowers Row*, 45; M'Vail, *Housing*, 43.
121. Seeley, 'Villages', 54.
122. Rymer, in *HW* 1, 9.
123. *Report of Whickham*, 6.
124. *Report of Commissioner*, 1850, 73.
125. *Miners' Advocate*, 21 Mar. 1874.
126. M'Vail, *Housing*, 38, 41.
127. *Report of Commissioner*, 1850, 58; Seeley, 'Villages', 304–5, appendix xiv.
128. Northumberland Miners' Mutual Confident Association, *Minutes*, Delegate Meeting, 19 June. 1875. Also Committee Meeting, 24 Apr. 1885 and 21 Nov. 1885.
129. E. Thorpe, 'Politics and Housing in a Durham Mining Town', *Mining and Social Change: Durham County in the Twentieth Century*, ed. M. Bulmer, London 1978, 110.
130. Bell, 'Reminiscences', 42; Moore, *Pit-Men*, 85.
131. Williams, *Derbyshire Miners*, 106. Also 'A.W.' to *Miners' Advocate*, 17 Jan. 1873.
132. *Colliery Guardian*, 26 Apr. 1878. Also National Library of Scotland, Lanarkshire Miners' County Union, *Minutes*, 25 May 1905.
133. Ridyard, *Abram*, 15; Duckham, *Scottish Industry*, 258.
134. Jones, 'Population', 295; *Report of Commissioner*, 1844, 41; Rymer, in *HW* 2, 30; T. Joseph to *Colliery Guardian*, 5 Sep. 1879.
135. Jones, 'Population', 297. See too *Report of Commissioner*, 1850, 48; *Miners' Advocate*, 1 Nov. 1873; *Colliery Guardian*, 15 Aug. 1879; Daunton, *Coal Metropolis*, 95.
136. Campbell, 'Slaves', 184.

137. Durland, *Fife*, 105.
138. Jevons, *Coal Trade*, 643.
139. Jevons, *Coal Trade*, 637; Morris and Williams, *South Wales*, 241–2.
140. Stonelake, 'autobiography', ch. 10, 2.
141. Raynes, *Conflicts*, 117.
142. *Report of Commissioner*, 1850, 13.
143. D. Rubinstein, *Victorian Homes*, Newton Abbot 1974, 127–8; Morris and Williams, *South Wales*, 240.
144. Neville, 'Yorkshire Miners', 808.
145. Jones, 'Population', 289; C. Evans, *Blaencwmdulais: A Short His-Tory of the Social & Industrial Development of Onllwyn and Banwen-Pyrddin*, Cardiff 1977, 83.
146. *Report of Commissioner*, 1850, 69.
147. Jevons, *Coal Trade*, 646–7.
148. Jones, 'Population', 298–301.
149. Jevons, *Coal Trade*, 647.
150. Daunton, *Metropolis*, 107, 114.
151. *Miner And Workman's Advocate*, 19 Dec. 1863.
152. *Report of Commissioner*, 1844, 58.
153. *Report of Commissioner*, 1850, 39.
154. *Miner & Workman's Advocate*, 19 Dec. 1863.
155. Evans interview.
156. Daunton, *Metropolis*, 112, 139–40.
157. *Select Committee Mines*, Q. 3, 5673–4, J. Normansell.
158. *Colliery Guardian*, 12 Mar. 1880.
159, Gregory, *Politics*, 55.
160. Hodges, 'I remember', 113.
161. Seeley, 'Villages', 57, 296.
162. Seeley, 'Villages', 295; 'Pro Zanto' to *Miners' Advocate*, 28 Mar. 1874.
163. J. W. White and R. Simpson, *Jubilee History of West Stanley Co-operative Society Limited. 1876 to 1926*, Pelaw-on-Tyne, 1926, 68–70, 224–5.
164. E. Lloyd, *History of the Crook and Neighbourhood Co-operative Corn Mill, Flour & Provision Society Limited*, Pelaw-on-Tyne 1916, 193–4.
165. Campbell, 'Slaves', 56.
166. Campbell, 'Slaves', 202.
167. Campbell, 'Slaves', 360.
168. M'Vail, *Housing*, 12.
169. Durland *Fife*, 108, 184.
170. M. Benney, *Charity Main*, 122, 163.

Chapter 5
THE MINER AT HOME
(pp. 112–141)

1. J. L. and B. Hammond, *The Skilled Labourer 1760–1832*, London 1919, 20.

2. *Report of Commissioner*, 1850, 9—10.
3. *Colliery Guardian*, 9 Aug. 1872.
4. Rymer, in *HW* 2, 21.
5. Williams, *Derbyshire Miners*, 58—9.
6. Challinor, *Lancashire Miners*, 245—6.
7. Bell, 'Reminiscences', 48.
8. Hodges, 'I remember', 21.
9. *Western Mail*, 13 Nov. 1913.
10. *Colliery Guardian*, 22 Aug. 1879. Also 5 Apr. 1878.
11. There was also a smaller repair and maintenance shift.
12. Wilson, *Autobiography*, 25. Also *Colliery Guardian*, 2 Jan. 1858; 7 Jan. 1910; McCormick and Williams, in *EHR* XII/2, 225; Hickling interview.
13. Halliday, *Just Ordinary*, 63.
14. Hair, 'Social History', 115—6.
15. *Colliery Guardian*, 3 July 1858; 27 Apr. 1900; Evison, 'Central Region', 182.
16. P. N. Jones, 'Workman's Trains in the South Wales Coalfield 1870—1926', *Transport History* 3/1 (Mar. 1970), 25.
17. Hair, 'Social History', 113; Redmayne, *Memories*, 18; *Reports of H.M. Inspectors of Mines*, A. H. Stokes, 1904; T. H. Corfe, R. W. Davies and A. Walton, *Historical Field Studies in the Durham Area: A Handbook for Teachers*, Durham 1967, 137.
18. 'A Man Knocking About' to *Miner & Workman's Advocate*, 24 Oct. 1863.
19. Hodges, 'I remember', 70.
20. Halliday, *Just Ordinary*, 64.
21. Rowe, *Wages*, 21.
22. Evison, 'Central Region', 158; Neville, 'Yorkshire Miners', 29 n.1.; *Colliery Guardian*, 7 Jan. 1910; Williams, *Derbyshire Miners*, 386.
23. *Colliery Guardian*, 7 Jan., 4 Mar. 1910.
24. Challinor, *Lancashire Miners*, 28, 117; Morris and Williams, *South Wales*, 260—2; Douglass, *Pit Talk*, 266.
25. Coombes, *Poor Hands*, 41.
26. Halliday, *Just Ordinary*, 63, 70, 72, 73. See also Durland, *Fife*, 62.
27. Halliday, *Just Ordinary*, 72—3. A unionist on afternoon shift could not attend meetings: Horner, *Rebel*, 24.
28. Jevons, *Coal Trade*, 617.
29. Challinor, *Lancashire Miners*, 117.
30. S. Chaplin, *Durham Mining Villages*, Durham 1971, 29.
31. Hair, 'Social History', 326; Challinor, *Lancashire Miners*, 69; *Select Committee Mines*, Q. 414, T. Burt; *Colliery Guardian*, 2 Jan. 1858; Hassan, 'Lothian', 227.
32. 'A Miner's Son' to *Miner & Workman's Advocate*, 12 Dec. 1863.
33. *Colliery Guardian*, 2 Jan. 1858; 'A Miner's Son' to *Miner & Workman's Advocate*, 12 Dec. 1863.
34. Stonelake 'autobiography', ch. 1, 12.

35. *Colliery Guardian*, 23 Mar., 8 Jun. 1900; Arnot, *South Wales*, 115.

36. Hair, 'Social History', 263.

37. *Colliery Guardian*, 24 Apr. 1858.

38. Hair, 'Social History', 326—9; T. Mountjoy to *Miners' Advocate*, 12 July 1873; *Report of Commissioner*, 1846, 131; Rymer, in *HW* 2, 25; J. Griffiths to *British Miner And General Newsman*, 24 Jan. 1863.

39. 'Old Chum' to *British Miner And General Newsman*, 24 Jan. 1863.

40. Reid, 'Saint Monday', 79.

41. Bell, 'Reminiscences', 27; M'Vail, *Housing*, 57.

42. Goodley, 'Bewicke Main', 9.

43. Hair, 'Social History', 259.

44. *Report of Commissioner*, 1857, 59; M'Vail, *Housing*, 44.

45. 'The Man Knocking About' to *Miner And Workman's Advocate*, 28 Nov. 1863.

46. Evison, 'Central Region', 80, 94; Seeley, 'Villages', 285.

47. M'Vail, *Housing*, 37—8.

48. Evison, 'Central Region', 399.

49. J. Lawson, 'The Influence of Peter Lee', *Mining and Social Change: Durham County in the Twentieth Century*, ed. M. Bulmer, London 1978, 104.

50. R. Page Arnot, *A History of the Scottish Miners*, London 1955, 39; Arnot, *The Miners*, 48; Williams, *Derbyshire Miners*, 217; T. Williams, *Digging for Britain*, London 1965, 10.

51. R. B. Outhwaite, 'Age at Marriage in England from the late Seventeenth to the Nineteenth Century', *Transactions of the Royal Historical Society* XXXIII (1973), 69. Also Mathias, *Industrial Nation*, 191; E. H. Phelps Brown, *The Growth of British Industrial Relations: A Study from the Standpoint of 1906—14*, London 1959, 7; S. G. Checkland, *The Rise of Industrial Society in England 1815—1885*, London 1964, 225.

52. Redmayne, *Memories*, 19. Also Hair, 'Social History', 253.

53. Outhwaite, in *TRHS* XXXIII, 58.

54. M. R. Haines, 'Fertility, Nuptiality, and Occupation: A Study of Coal Mining Populations and Regions in England and Wales in the Mid-Nineteenth Century', *Journal of Interdisciplinary History* VIII/2 (Autumn 1977), 259; Phelps Brown, *Industrial Relations*, 7.

55. Lee, *Home*, 38.

56. Hair, 'Social History', 284—9.

57. Benson, 'Compensation', 35 n.2; *Wolverhampton Chronicle*, 18 Aug. 1813; *Colliery Guardian*, 20 Feb. 1858; *Reports of H.M. Inspectors of Mines*, C. L. Morton, 1857, 134; Wakefield Public Library, Local Cuttings, 90/108.

58. Hair, 'Social History', Table 5. Also Evison, 'Central Region', 365—6; Campbell, 'Slaves', 180; Haines, in *JIH* VIII/2, 266.

59. Jones, 'Population', Figure 36. Also Hair, 'Social History', 77.

60. Hair, 'Social History', Table 5; Campbell, 'Slaves', 184; Durland,

Fife, 105; Jevons, *Coal Trade*, 643.

61. Durland, *Fife*, 78—9. 105—6.
62. Moore, *Pit-Men*, 67.
63. *Children's Employment Commission, Appendix to 1st Report*, 269, Rev. J. Blackburn.
64. Staffordshire County Record Office, D 593/n/2/2/31(b); D 593/N/3/2/16; Hodges, 'I remember', 72; Coombes, *Poor Hands*, 64; Bullock, *Bowers Row*, 59; M. Anderson, *Family Structure in Nineteenth Century Lancashire*, Cambridge 1971, 53, 148—9.
65. *Wolverhampton Chronicle*, 28 Feb. 1900.
66. Staffordshire County Record Office, D 593/N/2/2/31(b); D 593/N/3/2/1b; Evison, 'Central Region', 291, 293; Youngson Brown, 'Coal Industry', 256 n.1; Neville, 'Yorkshire Miners', 808—9; Jevons, *Coal Trade*, 129; Duckham, *Scottish Industry*, 257; Griffin, *Coalmining Industry*, 162; Durland, *Fife*, 107; F. J. Ball, 'Housing in an Industrial Colony: Ebbw Vale, 1778—1914', *The History of Working-Class Housing: A Symposium*, ed. S. D. Chapman, Newton Abbot, 1971.
67. Morris and Williams, *South Wales*, 242.
68. 'A Man Wheeling a Big Barrow' to *Miner & Workman's Advocate*, 26 Sep. 1863. Also Coombes, *Poor Hands*, 90—1; Durland, *Fife*, 122.
69. *Report of Whickham*, 13; M'Vail, *Housing*, 21, 35.
70. Brown interview; see also *Report of Commissioner*, 1859, 57.
71. M'Vail, *Housing*, 50—51.
72. Hodges, 'I remember', 81, 138.
73. Bell, 'Reminiscences', 59. Also Halliday, *Just Ordinary*, 16.
74. Hodges, 'I remember', 82.
75. Larmer interview. Also Bullock, *Bowers Row*, 45.
76. T. Wilson, *The Pitman's Pay and Other Poems*, Gateshead 1843. 149—50.
77. Holman, *Fife*, 100.
78. 'J.W.S.' to *Miners' Advocate*, 23 Aug. 1873. Also Holman, *Fife*, 81.
79. J. Wilson to *Miners' Advocate*, 24 May 1873.
80. Williams, *Derbyshire Miners*, 446.
81. Hair, 'Social History', 47; Hassan, 'Lothian', 238; Campbell, 'Slaves', 238—40; Burgess, *Industrial Relations*, 161.
82. Hair, 'Social History', 28, 33—5, 42—4, 55—6; Hassan, 'Lothian', 238, 246; Smith, *Economic Geography*, 126.
83. Rymer, in *HW* 1, 8.
84. E.g. *Colliery Guardian*, 18, 25 Jan., 15, 22 Feb., 12 July 1878; Taylor, 'Black Country', 161.
85. Neville, 'Yorkshire Miners', 810.
86. Storm-Clark, in *VS* XV/1, 64—5. Also Jones 'Population', 112, 131—3, 138.
87. John, 'Women', 72; Morris and Williams, *South Wales*, 237—8.
88. Campbell, 'Slaves', 241—4. Also Down and Warrington, *Somerset*, 38.

89. Jones, in *TH* 3/1, 22—3, 30.
90. Durland, *Fife*, 116, 118. Also Bullock, *Bowers Row*, 3, 30.
91. Halliday, *Just Ordinary*, 71—2. Also Bullock, *Bowers Row*, 185.
92. Hodges, 'I remember', 19.
93. Hodges, 'I remember', 20—21, 35, 74, 97. Also Durland, *Fife*, 118.
94. Anderson, *Family Structure*, 77.
95. Holman, *Fife*, 89, 100, 109; Bullock, *Bowers Row*, 12.
96. Coombes, *Poor Hands*, 21. Several commentators have stressed the contrast between the tawdry exteriors and the well-maintained interiors of colliery houses: Redmayne, *Memories*, 17—18; Hair, 'Social History', 268—70.
97. L. Oren, 'The Welfare of Women in Labouring Families: England, 1860—1950', *Clio's Consciousness Raised: New Perspectives on the History of Women*, eds M. S. Hartman and L. Banner, New York 1974, 226, 231, 239; Moore, *Pit-Men*, 147; Durland, *Fife*, 150; Coombes, *Poor Hands*, 100.
98. Campbell, 'Slaves', 283, 389.
99. Hodges, 'I remember', 45.
100. Evans interview; Jevons, *Coal Trade*, 70; John, 'Women', 181—2; *Select Committee Mines*, Q. 3,118—9, J. Normansell; Q. 4, 420, J. Booth; *Report of Commissioner*, 1844, 4; 1847, 44, 47, 49—52.
101. John, 'Women', 339. Dr John informs me that her recent work on Yorkshire census material confirms this point.
102. *Report of Commissioner*, 1844, 4. Tremenheere is of course making the point that the work is unsuitable for women.
103. Campbell, 'Slaves', 234—5. Also Hair, 'Social History', 264.
104. Hodges, 'I remember', 137.
105. E. Shorter, *The Making of the Modern Family*, Glasgow 1977, 74, 272. Also John, 'Women', 344.
106. Hodges, 'I remember', 19; Hair, 'Social History', 259; Evison, 'Central Region', 383.
107. Shorter, *Family*, 171; Anderson, *Family Structure*, 69, 76.
108. *Children's Employment Commission, 1st Report*, 162; *Appendix to 1st Report*, 22; Stonelake, 'autobiography', ch. 3, 8; Evison, 'Central Region', 383.
109. 'J.R.' to *Miners' Advocate*, 17 May 1873. Also Rymer, in *HW* 2, 19.
110. Evison, 'Central Region', 379, 382; *Colliery Guardian*, 31 Mar. 1905.
111. *Report of Whickham*, 23; Evison, 'Central Region', 390.
112. Holman, *Fife*, 105; Bullock, *Bowers Row*, 67. See also Anderson, *Family Structure*, 69.
113. Hodges, 'I remember', 8, 14, 134; Evans interview.
114. Evans interview; Stonelake, 'autobiography', ch. 3, 17; Hodges, 'I remember', 144; Bell, 'Reminiscences', 36—7; Bullock, *Bowers Row*, 73—5; Halliday, *Just Ordinary*, 29—31; Williams, *Digging*, 12.
115. Hodges, 'I remember', 6—7.

116. Hodges, 'I remember', 76; J. Lee, *This Great Journey: A Volume of Autobiography 1904–45*, London 1963, 34.
117. Stonelake, 'autobiography', ch. 2, 16.
118. Bell, 'Reminiscences', 54. Also Hodges, 'I remember', 61.
119. Anderson, *Family Structure*, 69–70.
120. *Children's Employment Commission, 1st Report*, 162.
121. Evison, 'Central Region', 391–2.
122. *Wolverhampton Chronicle*, 10 Jan. 1900.
123. Lee, *Journey*, 35; A. J. Taylor, *The Standard of Living in Britain in the Industrial Revolution*, London 1975, xxxv; Hodges, 'I remember', 105; Bullock, *Bowers Row*, 14.
124. *Colliery Guardian*, 18 Jan. 1878.
125. Hodges, 'I remember', 26.
126. Anderson, *Family Structure*, 78.
127. Hodges, 'I remember', 107.
128. Hodges, 'I remember', 46.
129. *Wolverhampton Chronicle*, 8 Jan. 1823.
130. Durham County Record Office, D/MRP 5/5(ii), G. McBurnie, 'Washington Old and New 1910–'.
131. *Report of Commissioner*, 1850, 26–7. Tremenheere's view accords with the impression gained from a reading of relevant autobiographical evidence.
132. Anderson, *Family Structure*, 108.
133. For dame schools see Seeley, 'Villages', 328; R. Colls, '"Oh Happy English Children" Coal, Class and Education in the North-East', *Past and Present* 73 (Nov. 1976), 90. For colliery schools see Colls, in *PP* 73; J. Benson, 'The Motives of 19th-Century English Colliery Owners in Promoting Day Schools', *Journal of Educational Administration and History* III/1 (Dec. 1970).
134. 'Cymno' to *Miner & Workman's Advocate*, 28 Nov. 1863. Also Bullock, *Bowers Row*, 109.
135. *Glamorgan Free Press*, 9 Oct. 1913.
136. Hodges, 'I remember', 99; Lee, *Journey*, 126.
137. Stonelake, 'autobiography', ch. 4, 3.
138. Hodges, 'I remember', 79.
139; Dalkin, 'Story'; Brown interview; McBurnie, 'Washington'; Bullock, *Bowers Row*, 163.
140. Stonelake, 'autobiography', ch. 4, 8.
141. Holman, *Fife*, 56. Also W. Paynter, *My Generation*, London 1972, 18.
142. *Report of Commissioner*, 1851, 4–5.
143. Hodges, 'I remember', 31.
144. Halliday, *Just Ordinary*, 23. Also Evans, *Seven Sisters*, 56–7; Bullock, *Bowers Row*, 11–12.
145. 'W.Y.C.' to *Miner & Workman's Advocate*, 12 Dec. 1863.
146. Williams, *Digging*, 11.
147. MacFarlane, 'Denaby Main', 127.
148. Durham County Record Office, N. Cowen, 'Of Mining Life and

Aal Its Ways', 2; also *Select Committee Mines*, Q. 18—19, T. Burt; Q. 2, 632—8, J. Ackersley; *Report of Commissioner*, 1846, 15, 45; Brown interview; Ridyard, *Abram*, 3; Stonelake, 'autobiography', ch. 4, 9.

149. Evison, 'Central Region', 371—2. Also John, 'Women', 121, 449—50; Bullock, *Bowers Row*, 47.

150. Evison, 'Central Region', 371, 373; John, 'Women', 337—8, 346; Brown interview; *British Miner And General Newsman*, 20 Sep. 1862; Durland, *Fife*, 58; Holman, *Fife*, 56; Bullock, *Bowers Row*, ix 47; P. Horn, *The Rise and Fall of the Victorian Servant*, Dublin 1975, 27.

151. Griffin, *East Midlands*, 112. Also Jevons, *Coal Trade*, 623—4; Mathias, *Industrial Nation*, 202; Bullock, *Bowers Row*, 184; *Select Committee Mines*, Q. 1, 192—3, 1, 511, W. Baxendale.

152. Parkinson, *True Stories*, 22—3.

153. Evans interview.

154. E. L. Trist and K. W. Bamforth, 'Some Social and Psychological Consequences of the Longwall Method of Coal-Getting', *Human Relations* 4/1 (1951), 34.

155. Challinor and Ripley, *Miners' Association*, 72.

156. Bullock, *Bowers Row*, 90.

157. Hodges, 'I remember', 97, 136. Also John, 'Women', 341—4; Challinor, *Lancashire Miners*, 159.

158. Anderson, *Family Structure*, 66—7.

159. John, 'Women', 344.

Chapter 6
THE MINER AT PLAY
(pp. 142—171)

1. B. Harrison, *Drink and the Victorians: The Temperance Question in England 1815—1872*, London 1971, 93.

2. *Report of Commissioner*, 1846, 36.

3. E. Pickering to J. J. Guest, 6 Mar. 1835, *Dowlais*, 68. Also W. R. Lambert, 'Drink and Work-Discipline in Industrial South Wales, c. 1800—1870', *Welsh History Review* VII/3 (June 1975), 296.

4. 'A Working Man at Thornley' to *Miner & Workman's Advocate* 27 June 1863. Also J. Gaven to *Miners' Advocate*, 17 Jan. 1874.

5. *Wolverhampton Chronicle*, 18 Nov. 1818.

6. Burgess, *Industrial Relations*, 169, 217; Challinor, *Lancashire Miners*, 109—10, 245.

7. *Wolverhampton Chronicle*, 29 Dec. 1824. Also 17 Mar. 1824; 19 Dec. 1858; 7 Mar. 1900.

8. *Report of Commissioner*, 1849, 14; 'Live and Let Live' to *Miner & Workman's Advocate*, 5 Dec. 1863; Seeley, 'Villages', 234; N. Longmate, *The Waterdrinkers: A History of Temperance*, London 1968, 14, 24.

9. Goodley, 'Bewicke Main', 13.

10. *Report of Commissioner*, 1844, 20.

11. K. Allan, 'The Recreations and Amusements of the Industrial Working Class, in the Second Quarter of the Nineteenth Century with Special Reference to Lancashire', (Manchester M.A. thesis, 1947), 60; *Western Mail*, 27 Feb. 1871; Harrison, *Drink*, 79; Longmate, *Waterdrinkers*, 23.

12. W. R. Lambert, 'The Welsh Sunday Closing Act, 1881', *Welsh History Review* VI/2 (Dec. 1972), 178.

13. Williams, *Derbyshire Miners*, 464.

14. Evison, 'Central Region', 435.

15. Williams, *Digging*, 23, 26.

16. J. Walvin, *Leisure and Society*, London 1978, 37.

17. Harrison, *Drink*, 328—9.

18. *Report of Commissioner* 1844, 20.

19. Harrison, *Drink*, 328—9; Williams, *Digging*, 23; Evans, *Seven Sisters*, 53—4.

20. Lambert, in *WHR* VII/3 289; Harrison, *Drink*, 37—9; M'Vail, *Housing*, 45.

21. E.g. *Colliery Guardian*, 9 Aug., 13 Sep. 1872; Evison, 'Central Region', 433; Morris and Williams, *South Wales*, 245—6.

22. Lambert, in *WHR* VI/2, 167.

23. Harrison, *Drink*, 47.

24. Bullock, *Bowers Row*, 167; *Wolverhampton Chronicle*, 15 Sep. 1819; Griffin, *East Midlands*, 44.

25. Challinor, *Lancashire Miners*, 246. Also John, 'Women', 345; J. Rowley, 'Drink and the Public House in Urban Nottingham, 1830—1860', *Bulletin of the Society for the Study of Labour History* 32 (Spring 1976), 7; T. Delves 'Popular Recreations and their Enemies', *Bulletin of the Society for the Study of Labour History* 32 (Spring 1976), 5.

26. Bullock, *Bowers Row*, 154—8. Also Parkinson, *True Stories*, 10.

27. Rymer, in *HW* 1, 6; *Wolverhampton Chronicle*, 8 Sep. 1824.

28. Raynes, *Conflicts*, xiv.

29. Williams, *Derbyshire Miners*, 59.

30. A. L. Lloyd, *Come all ye Bold Miners: Ballads and Songs of the Coalfields*, London 1952, 60.

31. *Report of Commissioner*, 1844, 18—19.

32. Hodges, 'I remember', 116.

33. Moore, *Pit-Men*, 140; Wilson, *Autobiography*, 26.

34. Lloyd, *Bold Miners*, 32—3.

35. Durland, *Fife*, 149. Also Stonelake, 'autobiography', ch. 2, 1; Hair, 'Social History', 281; M. Turner, 'The Miner's Search for Self-Improvement: The History of Evening Classes in the Rhondda Valley from 1862—1914', (Wales M.A. thesis, 1966), 44; Duckham, *Scottish Industry*, 260; Griffin, *Coalmining Industry*, 159—60. Compare R. Roberts, *The Classic Slum: Salford Life in the First Quarter of the Century*, Manchester 1971, 10, 96.

36. Horner, *Rebel*, 11. Also Bullock, *Bowers Row*, 151; T. Leckonby to *Miners' Advocate*, 22 Mar. 1873.

37. *Miners' Advocate*, 22 Nov. 1873. Also Holman, *Fife*, 76.
38. See for example Hair, 'Social History', 280—1; Bullock, *Bowers Row*, 148; Durland, *Fife*, 151; Horner, *Rebel*, 11.
39. Walvin, *Leisure*, 37. Also Hair, 'Social History', 256—7.
40. Wilson, *Pitman's Pay*, xiii—xiv. Also Hassan, 'Lothian', 305.
41. Holman, *Fife*, 71.
42. Harrison, *Drink*, 300—302, 319; Lambert, in *WHR* VII/3, 290; Walvin, *Leisure*, 11.
43. Hair, 'Social History', 392; Colls, *Rant*, 66.
44. Williams, *Derbyshire Miners*, 445; *Select Committee Mines*, Q. 7,125, A. McDonald; Ashton and Sykes, *Coal Industry*, 145—6.
45. *Report of Commissioner*, 1853, 31.
46. *Colliery Guardian*, 14 Sep. 1900.
47. Moore, *Pit-Men*, 83.
48. Moore, *Pit-Men*, 83. Also *British Miner And General Newsman*, 10 Jan. 1863; *Hamilton Advertiser*, 22 Dec. 1877.
49. Durland, *Fife*, 132—3.
50. *Report of Commissioner*, 1844, 41. Also 1853, 28; Hassan, 'Lothian', 308.
51. *Report of Commissioner*, 1850, 50; 1851, 30. Also Hair, 'Social History', 413.
52. *Report of Commissioner*, 1850, 55; *Colliery Guardian*, 20 Nov. 1858; *British Miner & General Newsman*, 6 Dec. 1862; Seeley, 'Villages', 131, 230, 234, 238; Hassan, 'Lothian', 308; Neville, 'Yorkshire Miners', 815.
53. Pollard, *Genesis*.
54. Evans, *Seven Sisters*, 60; Seeley, 'Villages', 131; Colls, *Rant*, 93.
55. *Report of Commissioner*, 1850, 50—51.
56. *Report of Commissioner*, 1853, 28.
57. 'A Member of the Institute' to *British Miner & General Newsman*, 29 Nov. 1862.
58. Turner, 'Self-Improvement', 166. Also H. Francis, 'The Origins of the South Wales Miners' Library', *History Workshop* 2 (Autumn 1976).
59. *Colliery Guardian*, 4 Sep. 1858; Seeley, 'Villages', 186, 282—3. But see M. Tylecote, *The Mechanics' Institutes of Lancashire and Yorkshire before 1851*, Manchester 1957, 108.
60. Challinor, *Lancashire Miners*, 179.
61. *Select Committee Mines*, Q. 9, 660, Rev. H. R. Sandford; Turner, 'Self-Improvement'; *Colliery Guardian*, 28 May 1880; Redmayne, *Memories*, 37; Ridyard, *Abram*, 11; Neville, 'Yorkshire Miners', 825—7.
62. *Sheffield Independent*, 8 May 1893. Also *Wigan Observer*, 20 Oct. 1894; *Dean Forest Mercury*, 20 Oct. 1893; *Newcastle Weekly Chronicle*, 11 May 1889; *Derbyshire Times*, 8 Sep. 1883.
63. *Colliery Guardian*, 4 Mar. 1910.
64. R. W. Malcomson, *Popular Recreations in English Society 1700—*

1850, Cambridge 1973, 119; also 49, 122–3; Redmayne, *Memories*, 31; *Wolverhampton Chronicle*, 29 July 1818.
65. *Wolverhampton Chronicle*, 8 Nov. 1837.
66. *Wolverhampton Chronicle*, 11 Oct., 29 Nov. 1815; 7 Nov. 1838; Malcolmson, *Recreation*, 118–9, 124–6; *The Miner*, 18 Apr. 1863; Hassan, 'Lothian', 294; J. Mott, 'Miners, Weavers and Pigeon Racing', *Leisure and Society in Britain*, eds M. A. Smith, S. Parker and C. S. Smith, London 1973, 86.
67. *Colliery Guardian*, 25 Sep. 1858. Also *Miner & Workman's Advocate*, 29 Aug. 1863.
68. Neville, 'Yorkshire Miners', 821.
69. Williams, *Derbyshire Miners*, 465.
70. Stonelake, 'autobiography', ch. 13, 1. Also Parkinson, *True Stories*, 63, 73.
71. Bullock, *Bowers Row*, 80.
72. *Report of Commissioner*, 1844, 45; also 50, 58; Hair, 'Social History', 277–8.
73. *Report of Commissioner*, 1859, 56; *Colliery Guardian*, 13 June 1879; Holman, *Fife*, 79; Seeley, 'Villages', 287; Griffin, *Coalmining Industry*, 159.
74. Halliday, *Just Ordinary*, 17. Also Evans interview; Brown interview; Bullock, *Bowers Row*, 14–16.
75. *Select Committee Mines*, Q. 428, T. Burt; Seeley, 'Villages', 176, 279; Hair, 'Social History', 277–8; Griffin, *East Midlands*, 51.
76. Holman, *Fife*, 44–5.
77. *Colliery Guardian*, 26 July 1872.
78. Hodges, 'I remember', 9; Holman, *Fife*, 46.
79. *Colliery Guardian*, 4 June 1880.
80. Colls, *Rant*, 52–3, 143; Bullock, *Bowers Row*, 53–4; Lewis, *Coal Mining*, 30; Durland, *Fife*, 124; Walvin, *Leisure*, 97–8; MacFarlane, 'Denaby Main', 117.
81. Stonelake, 'autobiography', ch. 5, 13–14. Also Evans interview.
82. Evans interview; Holman, *Fife*, 40; Walvin, *Leisure*, 103–4.
83. *British Miner And General Newsman*, 20 Sept. 1862.
84. Evans, *Seven Sisters*, 133, 135.
85. Hair, 'Social History', 278.
86. J. F. Russell and J. H. Elliot, *The Brass Band Movement*, London 1936, 180.
87. *Report of Commissioner*, 1844, 40; Stonelake, 'autobiography', ch. 3, 12; Lee, *Journey*, 21.
88. Wilson, *Autobiography*, 23.
89. Jevons, *Coal Trade*, 629.
90. Evans, *Seven Sisters*, 145.
91. Bullock, *Bowers Row*, 82–3. For a similar reaction see Roberts, *Classic Slum*, 140–1.
92. Duckham, *Scottish Industry*, 292; Jevons, *Coal Trade*, 635; Neville, 'Yorkshire Miners', 820; Durham County Record Office,

transcript of interview with Mr and Mrs David and Ruby Larmer and Rose, 2 May 1976.

93. *Colliery Guardian*, 12 Sep. 1879; also 5 Sep. 1879; *Wolverhampton Chronicle*, 23 Aug. 1837.

94. *Wolverhampton Chronicle*, 9 May 1821; Rymer, in *HW* 1, 5; Hair, 'Social History', 276; Wilson, *Autobiography*, 68.

95. Bullock, *Bowers Row*, 79. Also D. Larmer interview.

96. Bullock, *Bowers Row*, 158; *Miner And Workman's Advocate*, 31 Oct. 1863; 'Veritas' to *British Miner & General Newsman*, 18 Oct. 1862.

97. 'A Marley-Hill Miner' to *British Miner And General Newsman*, 14 Feb. 1863.

98. D. Larmer interview.

99. Walvin, *Leisure*, 10, 87; Malcolmson, *Recreation*, 171.

100. D. Larmer interview; Redmayne, *Memories*, 91; *Colliery Guardian*, 26 Apr. 1878; Ridyard, *Abram*, 32.

101. *Colliery Guardian*, 4 July 1973.

102. D. Larmer interview. This section is based largely on Mott, 'Pigeon'. See also Redmayne, *Memories*, 9.

103. Bullock, *Bowers Row*, 84.

104. Evans *Seven Sisters*, 149.

105. Evans interview.

106. MacFarlane, 'Denaby Main', 116.

107. Bullock, *Bowers Row*, 78. Also Lawson, *Home*, 79.

108. Coombes, *Poor Hands*, 141—6. Also *Colliery Guardian*, 28 Aug. 1858; Evans interview; Bullock, *Bowers Row*, 78—80.

109. Parkinson, *True Stories*, 9.

110. I owe this phrase to B. Trinder, *The Industrial Revolution in Shropshire*, Chichester, 1973, 285.

111. Halliday, *Just Ordinary*, 13.

112. *Hamilton Advertiser*, 27 Oct. 1877.

113. Gregory, *Politics*, 7, 92—3.

114. Gregory, *Politics*, 93 n.1. Also Hodges, 'I remember', 10.

115. Hassan, 'Lothian', 304; *Wolverhampton Chonicle*, 7 Dec. 1825.

116. Hair, 'Social History', 291—4; John 'Women', 346; Trinder, *Shropshire*, 284.

117. Griffin, *East Midlands*, 45.

118. Morris and Williams, *South Wales*, 246. Also Arnot, *South Wales*, 166.

119. *British Miner And General Newsman*, 18 Oct. 1862.

120. See Moore, *Pit-Men*, 5—14.

121. Hair, 'Social History', 289, 416.

122. Griffin, *East Midlands*, 45; Hair, 'Social History', 293.

123. Holman, *Fife*, 74—5.

124. Bullock, *Bowers Row*, 128.

125. T. Leckonby to *Miners' Advocate*, 29 Mar. 1873.

126. Parkinson, *True Stories*, 96.

127. Hair 'Social History', 28 n.1, 416—7.

128. Bullock, *Bowers Row*, 136. Also Parkinson, *True Stories*, 13; Lewis, *Coal Mining*, 36.

129. Williams, *Derbyshire Miners*, 78.

130. Moore, *Pit-Men*, 68, 70.

131. Paynter, *My Generation*, 27.

132. J. Foster, *Class Struggle and the Industrial Revolution: Early Industrial Capitalism in three English towns*, London 1974, 223.

133. Parkinson, *True Stories*, 9.

134. Halliday, *Just Ordinary*, 37—8; Hodges, 'I remember', 10, 12, 30; Neville, Yorkshire Miners', 829.

135. Bullock, *Bowers Row*, 136—40. Also Moore, *Pit-Men*, 129; Jones, 'Population', 10, 11, 15.

Chapter 7
THE MINER AND HIS CLUBS
(pp. 172—213)

1. Williams, *Derbyshire Miners*, 60.

2. Hammond, *Skilled Labourer*, 20. Also *Post Magazine and Insurance Monitor*, 31 Jan. 1885; *Nuneaton Chronicle*, 27 Apr. 1888; *Staffordshire Sentinel*, 12 Apr. 1890; *Mining Journal*, 24 Oct. 1857; *Colliery Guardian*, 20 Jan. 1866.

3. Williams, *Derbyshire Miners*, 58.

4. *Colliery Guardian*, 5 Sep. 1873.

5. *Colliery Guardian*, 9 Aug. 1872; 25 Oct. 1872.

6. 'Poor Married Man' to *Miners' Advocate*, 27 Oct. 1873. Also Rymer, in *HW* 2, 21; P. Pearce to *Miner and Workmen's Examiner*, 21 Apr. 1876.

7. Nor did the legal action (or the threat of it) prove of much assistance to those injured and bereaved by colliery accidents. See Benson, 'Compensation', 91—135; Morris and Williams, *South Wales*, 195, 198—200; *Colliery Guardian*, 21 Apr. 1905; *Mexboro' & Swinton Times*, 24 July 1903; *Sheffield Independent*, 17 Apr. 1903; 16 Jan. 1904; 7 July 1906; 23 Apr. 1909.

8. Williams, *Derbyshire Miners*, 58, 60.

9. Challinor, *Lancashire Miners*, 162.

10. Dennis, Henriques and Slaughter, *Coal*, 132—4.

11. Benson, 'Compensation', 38—59; *Wolverhampton Chronicle*, 11 Aug. 1824; *Children's Employment Commission, 1st Report*, 81, 162, 233, 235; Arnot, *The Miners*, 49.

12. *Colliery Guardian*, 11 Jan., 24 May 1878; 11 July 1879.

13. *Colliery Guardian*, 30 Apr. 1880.

14. *Colliery Guardian*, 9 Feb. 1900. See also *Western Mail* 21 Nov. 1913.

15. *Report of Commissioner*, 1844, 49.

16. *Royal Commission on the Poor Laws and Relief of Distress*, 740. Also Appendix, 126.

17. *Barnsley Chronicle*, 26 Jan., 9 Feb. 1867. Also Benson, 'Compensation', 46—7; *Children's Employment Commission*, Appendix to *1st Report*, 234.

18. *Wolverhampton Chronicle*, 11 Feb. 1880. Also Benson, 'Compensation', 45.

19. M. E. Rose, 'The Allowance System under the New Poor Law', *Economic History Review* XIX/3 (1966), 620.

20. 'W.H.' to *Wigan Observer*, 29 Sep. 1871. Also Benson, 'Compensation', 41, 82–7; Holman, *Fife*, 90.

21. Benson, 'Compensation', 49–54.

22. *The Provident*, Apr. 1881.

23. *Dean Forest Mercury*, 1 July 1887.

24. *Barnsley Chronicle*, 19 Oct. 1889; *Barnsley Independent*, 19 Oct. 1889.

25. *Wolverhampton Chronicle*, 28 Aug., 4 Sep., 25 Dec. 1816; 31 Jan., 7 Feb. 1838; Williams, *Derbyshire Miners*, 58, 282, 330, 425, 454; Moore, *Pit-Men*, 143; *Hamilton Advertiser*, 26 Feb. 1887.

26. *Colliery Guardian*, 25 Jan. 1878.

27. *Western Mail*, 5 Feb. 1879.

28. *Colliery Guardian*, 27 Nov. 1858.

29. Challinor, *Lancashire Miners*, 125.

30. *Wolverhampton Chronicle*, 3, 17 July 1816; *Colliery Guardian*, 2 Jan. 1858; 27 June 1879.

31. *Colliery Guardian*, 7 Aug., 4 Sep. 1858; 11 June 1880. Also *Wolverhampton Chronicle*, 10 Apr. 1822; Challinor, *Lancashire Miners*, 125; Arnot, *Scottish Miners*, 42–3.

32. *Colliery Guardian*, 11 May 1861; J. W. Fletcher and R. Chapman to *Wigan Observer*, 26 Oct. 1895.

33. This section is based largely on my article, 'Colliery Disaster Funds, 1860–1897', *International Review of Social History* XIX/1 (1974).

34. Benson, 'Compensation', 199; Raynes, *Conflicts*, 133; *Colliery Guardian*, 22 Mar. 1878.

35. *Western Mail*, 12 Nov. 1913.

36. *Colliery Guardian*, 24 May 1878; Ridyard, *Abram*, 16; Williams, *Derbyshire Miners*, 330.

37. *Mining Journal*, 10 Aug. 1850; *Colliery Guardian*, 11 July 1879; 21 Jan. 1910; Hassan, 'Lothian', 301; Gregory, *Politics*, 166 n.2; Williams, *Derbyshire Miners*, 444.

38. Benson, 'Compensation', 216–28; *Children's Employment Commission, 1st Report*, 609, 624.

39. Mee, *Enterprise*, 139–40; Machin, *Yorkshire*, 28–9.

40. Full details of English coalfields in the second half of the nineteenth century will be found in Benson, 'Compensation', 216–61. Also Hair, 'Social History', 393–4; *Children's Employment Commission*, Appendix to *1st Report*, 1, 68, 304; *Colliery Guardian*, 13 Sep. 1872.

41. Benson, 'Compensation', 257.

42. *Reports of H.M. Inspectors of Mines*, C. Morton, 1859, 64.

43. *Barnsley Chronicle*, 20 July 1867.

44. Northumberland Miners' Mutual Confident Association, *Minutes*,

21 Nov. 1891.

45. *Children's Employment Commission,* Appendix to *1st Report,* 162.

46. R. Fynes, *The Miners of Northumberland and Durham,* Sunderland 1873. 140. Also Benson, 'Compensation', 226—7.

47. *Children's Employment Commission, 1st Report,* 13, 88, 93—4, 112—3, 645, 663—4; *2nd Report,* 19, 21, 42, 148, 272, 299, 302, 310, 312, 323, 326, 337, 795—6, 843; *Report of Commissioner,* 1846, 18; 1849, 8; *Colliery Guardian,* 24 Aug. 1900; Morris and Williams, *Industry,* 243.

48. *Children's Employment Commission,* Appendix to *1st Report,* 312.

49. *Children's Employment Commission,* Appendix to *1st Report,* 337.

50. *Report of Commissioner,* 1844, 49.

51. Benson, 'Compensation', 289—93.

52. Benson, 'Compensation', 265—8; Morris and Williams, *South Wales,* 243.

53. Benson, 'Compensation', 268—9; Hair, 'Social History', 395; *Colliery Guardian,* 3 Apr. 1858.

54. 'One That is not Asleep' to *Miner & Workman's Advocate,* 18 Mar. 1865.

55. *Transactions and Results of the National Association of Coal, Lime and Iron-stone Miners . . .,* London 1864, xi.

56. Benson, 'Compensation', 285—7.

57. *Colliery Guardian,* 14 Jan. 1870.

58. *Yorkshire Post,* 27 Dec. 1866; *Wigan Observer,* 27 June 1885; *Hamilton Advertiser,* 22 Dec. 1877; *Western Mail,* 7 Nov. 1913.

59. Benson, 'Compensation', 371; *Select Committee on Accidents in Coal Mines, 2nd Report,* 1854, 27; F. McKenna, 'Victorian Railway Workers', *History Workshop* 1 (Spring 1976), 35—6.

60. *Labour Tribune,* 24 July 1886. Also Benson, 'Compensation', 298—300.

61. 'A Workman' to *Derbyshire Times,* 7 Aug. 1880. Also 'One Interested' to *Derbyshire Courier,* 11 Oct. 1879.

62. 'A Man Behind the Door at Coxhoe' to *Miner & Workman's Advocate,* 3 Oct. 1863.

63. Benson, 'Compensation', 301—3; Hair, 'Social History', 395.

64. *Children's Employment Commission, 1st Report,* 546.

65. W. Edwards to *Miner & Workman's Advocate,* 15 Aug. 1863.

66. 'A Miner at Coxhoe' to *Miner & Workman's Advocate,* 24 Oct. 1863; 'A Man Behind the Door at Coxhoe' to *Miner & Workman's Advocate,* 31 Oct. 1863.

67. *British Miner,* 24 Jan. 1863.

68. W. Mason to *Miner & Workman's Advocate,* 28 May 1864. Also *Colliery Guardian,* 22 Dec. 1871.

69. Williams, *Digging,* 16.

70. J. Benson, 'The Establishment of the West Riding of Yorkshire

Miners' Permanent Relief Fund', *Studies in the Yorkshire Coal Industry*, eds J. Benson and R. G. Neville, Manchester 1976, 93–4.
71. Moore, *Pit-Men*, 143; *Hamilton Advertiser*, 19 Mar. 1887.
72. Hair, 'Social History', 411; H. O. Horne, *A History of Savings Banks*, Oxford 1947, 187.
73. Walvin, *Leisure*, 61.
74. Durham County Record Office, D/DCB, Ledger of City and County of Durham Permanent Benefit Building Society; *Colliery Guardian*, 12 Mar. 1880.
75. Horne, *Banks*, 85; *Report of Commissioner*, 1847, 27.
76. Horne, *Banks*, 98.
77. *Colliery Guardian*, 4 July 1873. Also Hassan, 'Lothian', 307.
78. Evison, 'Central Region', 432; Horne, *Banks*, 188. Also *Colliery Guardian*, 2 May 1879.
79. Ridyard, *Abram*, 3. Also Durland, *Fife*, 22.
80. *Children's Employment Commission*, Appendix to *1st Report*, 64; Hair, 'Social History', 256.
81. *Children's Employment Commission*, Appendix to *1st Report*, 64, 114; Wilson *Autobiography*, 29.
82. Evison, 'Central Region', 232. Also Hair, 'Social History', 403.
83. *Children's Employment Commission*, Appendix to *1st Report*, 61. Also 67; *2nd Report*, 45; John, 'Women', 445; Evans, *Seven Sisters*, 138.
84. *Report of Commissioner*, 1854, 14–15. Also *Children's Employment Commission*, Appendix to *1st Report*, 24, 835; Hassan, 'Lothian', 306; Wilson, 'Miners', 93.
85. *Colliery Guardian*, 4 June 1880. Also Benson, 'Compensation', 429–34, 437, 439.
86. *Wigan Observer*, 26 Oct. 1895; *Barnsley Chronicle*, 1 Aug. 1896.
87. The following section is based essentially on my article, 'The Thrift of English Coal-Miners, 1860–95' *Economic History Review* XXXI/3 (Aug. 1978) where full details will be found. See also *Hamilton Advertiser*, 24 Nov. 1877; 23 Apr. 1887; *Western Mail*, 27 Feb. 1871; Wilson, 'Miners', 325.
88. *Children's Employment Commission*, Appendix to *1st Report*, 303.
89. *Labour Tribune*, 3 May 1890.
90. *Children's Employment Commission*, Appendix to *1st Report*, 75.
91. Durland, *Fife*, 78. Also Holman, *Fife*, 85.
92. *Barnsley Chronicle*, 10 Mar. 1877; 10 Mar. 1883; 9 Mar. 1895.
93. *Colliery Guardian*, 27 Apr. 1861; *Labour Tribune*, 14 Jan. 1888.
94. *Colliery Guardian*, 25 Oct. 1872.
95. Roberts, *Classic Slum*, 64. Also Anderson, *Family Structure*, 147.
96. B. E. Supple, *The Royal Exchange Assurance: A History of Insurance, 1720–1970*, Cambridge 1970, 113, 129; D. Morrah, *A History of Industrial Life Assurance*, London 1955, 9; 'M.H.' to *Miner & Workman's Advocate*, 19 Jan. 1861; *Newcastle Weekly Chronicle*, 9 Sep. 1865.

97. Benson, in *EHR* XXXI/3, 414—5.
98. Bullock, *Bowers Row,* 169.
99. *Accrington Times,* 17 Nov. 1883; *Bolton Journal,* 27 June 1885; *Miner and Workmen's Examiner,* 4 Aug. 1876; *Colliery Guardian,* 23 June 1905.
100. *Independent Order of Odd-Fellows, Manchester Unity Friendly Society. Supplementary Report, July 1st, 1872,* 98; J. Payne to *Barnsley Chronicle,* 12 Jan. 1867; Benson 'Compensation', Table I; *Children's Employment Commission,* Appendix to *1st Report,* 151, 204—5.
101. Benson, 'Compensation', 400—28, 440—1.
102. *The Provident,* 15 Feb. 1884.
103. Northumberland and Durham Miners' Permanent Relief Fund, *10th Annual Report,* 9. Also *Derbyshire Times,* 4 Oct. 1879.
104. Benson, in *EHR* XXXI/3, 415—6.
105. West Riding Miners' Permanent Relief Fund, *24th Annual Report,* 13.
106. *Colliery Guardian* 10 May 1878; Taylor, 'Black Country', 118; Neville, 'Yorkshire Miners', 52; Burgess, *'Industrial Relations,* 184; Challinor, *Lancashire Miners,* 49; Campbell, 'Slaves', 363.
107. Griffin, *Coalmining Industry,* 87—8. Also Youngson Brown, 'Coal Industry 206, 210; Neville, 'Yorkshire Miners', 64; Taylor, 'Black Country' 30; Burgess, *Industrial Relations,* 200.
108. Northumberland Miners' Mutual Confident Association, *Minutes,* Delegate Meeting, 18 Dec. 1875; *Monthly Circular,* May 1904; Rowe, *Wages,* 36; Campbell, 'Slaves', 233, 437; Evison, 'Central Region', 212—3; 'A Lover of Fair Play' to *Wigan Observer,* 7 Mar. 1873; 'A Miner' to *Miners' Advocate,* 31 Jan. 1874.
109. 'An Eston Miner' to *Miners' Advocate,* 14 Mar. 1873.
110. Davison, *Northumberland Miners,* 125.
111. Challinor, *Lancashire Miners,* 199.
112. Coombes, *Poor Hands,* 166—7.
113. *Colliery Guardian,* 18 July, 29 Aug. 1879. Also 1 Aug. 1879; Youngson Brown, 'Coal Industry', 196.
114. Lanarkshire Miners' County Union, *Minutes,* Special Conferences, 18 Mar., 29 Apr. 1902; South Wales Miners' Federation, Council Meeting, 17 July 1908; Jevons, *Coal Trade,* 124.
115. Lanarkshire Miners' County Union, *Minutes,* Council Meeting, 29 Apr. 1907; *Colliery Guardian,* 16 June 1905. Cf. Arnot, *South Wales,* 322.
116. Wilson, 'Miners', 324; Campbell, 'Slaves', 124, 147.
117. Youngson Brown, 'Coal Industry', 10.
118. Campbell, 'Slaves', 268, 288, 299; Hassan, 'Lothian', 269—71.
119. Campbell, 'Slaves', 401. Also 294.
120. Mee, *Enterprise,* 183. See too Wilson, 'Miners', 102; Hassan, 'Lothian', 303.
121. Campbell, 'Slaves', 109, 130; Challinor, *Lancashire Miners,* 206; Williams, *Derbyshire Miners,* 72; Neville, Yorkshire Miners', 309—12.

122. Challinor, *Lancashire Miners*, 80.
123. Taylor, 'Black Country', 177.
124. MacFarlane, 'Denaby Main', 141–2.
125. Neville, 'Yorkshire Miners', 94–6.
126. Morris and Williams, *South Wales*, 276–7; Hassan, 'Lothian', 171–2; J. H. Morris and L. J. Williams, 'The Discharge Note in the South Wales Coal Industry, 1841–1898', *Economic History Review* X/2 (1957), 288–9.
127. *Colliery Guardian*, 1 Mar. 1889; Burgess, *Industrial Relations*, 199.
128. *Miners' Advocate*, 12 Apr. 1873.
129. Wilson, *Autobiography*, 48.
130. Parfitt, *Somerset Miner*, 29; Horner, *Rebel*, 24; Challinor, *Lancashire Miners*, 19.
131. Bullock, *Bowers Row*, 139.
132. Coombes, *Poor Hands*, 50–1.
133. J. Peile to J. Guest, 30 Aug. 1831, *Dowlais*, 58.
134. Evison, 'Central Region', 201.
135. *Miner & Workman's Advocate*, 12 Dec. 1863.
136. *Colliery Guardian*, 10 Apr. 1879.
137. *Colliery Guardian*, 4 June 1880.
138. MacFarlane, 'Denaby Main', 115–6. Also Morris and Williams, *South Wales*, 262–4; Campbell, 'Slaves', 110–11; K. O. Morgan, *Keir Hardie Radical and Socialist*, London 1975, 117.
139. Evans, *Seven Sisters*, 73–4.
140. Hair, 'Social History', 376–7; Challinor and Ripley, *Miners' Association*, 18; Burgess, *Industrial Relations*, 172.
141. Challinor, *Lancashire Miners*, 25–8.
142. Fynes, *Miners*, 36; S. Webb, *The Story of the Durham Miners (1662–1921)*, London 1921, 31; Griffin, *Coalmining Industry*, 87–8.
143. Challinor and Ripley, *Miners' Association*, 57–8; Griffin, *Coalmining Industry*, 88.
144. Fynes, *Miners*, 36. Also Challinor, *Lancashire Miners*, 31–3; Hassan, 'Lothian', 255.
145. Campbell, 'Slaves', 247, 356; Challinor, *Lancashire Miners*, 16–17; Burgess, *Industrial Relations*, 177, 184; Youngson Brown, 'Coal Industry', 212; Williams, *Digging*, 41.
146. H. A. Clegg, A. Fox and A. F. Thompson, *A History of British Trade Unions since 1889* I, Oxford 1964, 103; Ridyard, *Abram*, 13; *9th Report of the Labour Correspondent*, 18; Arnot, *South Wales*, 184 n.3, 317.
147. *Miners' Advocate*, 5 Apr. 1873; Lancashire and Cheshire Miners' Federation, *Minutes*, 15 Feb. 1896; Burgess, *Industrial Relations*, 211.
148. Neville, 'Yorkshire Miners', 416–7.
149. Williams, *Derbyshire Miners*, 256, 433; *Sheffield Daily Telegraph*, 31 Aug. 1896; South Wales Miners' Federation, *Minutes*, Council Meeting, 20 Dec. 1909.

150. National Union of Mineworkers (North Wales Area), Point of Ayr Lodge, Minutes, Committee, 2 Mar. 1894.
151. *Wolverhampton Chronicle*, 31 Jan. 1900; Lanarkshire Miners' County Union, *Report*, 31 Dec. 1907, 7.
152. Williams, *Derbyshire Miners*, 433.
153. Campbell, 'Slaves', 387; Challinor, *Lancashire Miners*, 81.
154. Benson, 'Compensation', 350—1; *Potts' Register 1889*, 217; Arnot, *South Wales*, 55.
155. *Miners' Advocate*, 27 Sep. 1873.
156. *Colliery Guardian*, 27 Sep. 1872.
157. Benson, 'Compensation', 351—2.
158. *Colliery Guardian*, 21 June 1878.
159. Williams, *Derbyshire Miners*, 129, 165.
160. *Colliery Guardian*, 25 Apr. 1879; Youngson Brown, 'Coal Industry', 208.
161. Clapham, *Economic History*, 164.
162. Clapham, *Economic History*, 164; Northumberland and Durham Miners' Permanent Relief Fund, *23rd Annual Report*, 7.
163. Benson, 'Compensation', 352—3; Burgess, *Industrial Relations*, 210; Arnot, *Scottish Miners*, 70; Morgan, *Keir Hardie*, 42.
164. Arnot, *South Wales*, 61; S. and B. Webb, *The History of Trade Unionism*, London 1894, 750.
165. J. E. Morgan, *A Village Workers Council and What it Accomplished: Being a Short History of the Lady Windsor Lodge, South Wales Miners' Federation*, Pontypridd 1956, 15—16. Also *Colliery Guardian*, 4 May, 8 June 1900; 6 Jan., 20 June 1905; South Wales Miners' Federation, *Minutes*, Council Meeting, 27 Apr. 1908.
166. Webbs, *Trade Unionism*, 750.
167. Williams, *Derbyshire Miners*, 453.
168. Benson, 'Compensation', 112—6.
169. *Colliery Guardian*, 14 Jan. 1910; Rowe, *Wages*, 38.
170. Benson, 'Compensation', 149.
171. *Miners' Advocate*, 17 Jan., 7 Feb. 1874.
172. Benson, 'Compensation', 407. See also 408—10; *Liverpool Mercury*, 24 Mar. 1891.
173. Northumberland Miners' Mutual Confident Association, *Minutes*, Delegate Meeting, 22 June 1875.
174. Northumberland Miners' Mutual Confident Association, *Minutes*, Committee Meetings, 1 Apr. 1882; 26 Mar. 1886.
175. Point of Ayr Lodge, Committee Meeting, 7 June 1895.
176. Williams, *Derbyshire Miners*, 241—2.
177. *Colliery Guardian*, 24 Jan. 1879.
178. Northumberland Miners' Mutual Confident Association, *Minutes*, Emigration Rules, 18 July 1881. The scheme ended in 1883.
179. Hassan, 'Lothian', 306—7; Williams, *Derbyshire Miners*, 137; J. Harris, *William Beveridge: A Biography*, Oxford 1977, 101.
180. Williams, *Derbyshire Miners*, 243.

181. Lanarkshire Miners' County Union, *Minutes*, Council Meeting, 31 July 1902; Williams, *Derbyshire Miners*, 463, 541.
182. This section is based largely on Benson, in *EHR* XXXI/3; J. Benson, 'English Coal-Miners' Trade-Union Accident Funds, 1850—1900', *Economic History Review* XXVIII/3 (Aug. 1975). For North Wales see e.g. *Colliery Guardian*, 14 Apr. 1905; *Miner & Workman's Advocate*, 5 Sep. 1863. For Scotland see *Miners' Advocate*, 31 Jan., 21 Feb. 1874; *The Miner*, 23 May 1863; *Colliery Guardian*, 11 Feb. 1910; Hassan, 'Lothian', 307—7; Campbell, 'Slaves', 380.
183. Bulman, *Coal Mining*, 44.
184. Hair, 'Social History', 424.
185. Hair, 'Social History', 380 n.1; Arnot, *The Miners*, 408; Moore, *Pit-Men;* Morris and Williams, *South Wales*, 253—4.
186. Northumberland Miners' Mutual Confident Association, *Minutes*, Committee Meeting, 8 Sep. 1882.
187. Morris and Williams, *South Wales*, 270.
188. R. G. Neville, 'In the Wake of Taff Vale: The Denaby and Cadeby Miners' Strike and Conspiracy Case, 1902—06', *Studies in the Yorkshire Coal Industry*, eds J. Benson and R. G. Neville, Manchester 1976, 151.
189. Williams, *Derbyshire Miners*, 137.
190. *North Wales Miners' Magazine*, July 1903; Williams, *Derbyshire Miners*, 241.
191. Williams, *Derbyshire Miners*, 113; Arnot, *The Miners*, 260; Lanarkshire Miners' County Union, *Report*, 31 Dec. 1908, 3.
192. Williams, *Digging*, 19.
193. D. H. Lawrence, *The Complete Short Stories* I, London 1955, 53.
194. Morgan, *Council*, 25.
195. Benson, in *BSSLH*, 32, 22.
196. Webbs, *Trade Unionism*, 298; Hassan, 'Lothian', 264; Gregory, *Politics*, 8; Burgess, *Industrial Relations*, 199; Moore, *Pit-Men*, 39.
197. Coombes, *Poor Hands*, 134—5.
198. Gregory, *Politics*, 8.
199. 'A Miner' to *Miners' Advocate*, 1 Nov. 1873.
200. Williams, *Digging*, 26.
201. *Miners' Advocate*, 7 June 1873.
202. *Miners' Advocate*, 27 Sep. 1873.
203. Northumberland Miners' Mutual Confident Association, *Minutes*, Committee Meeting, 4 Apr. 1885.
204. Northumberland Miners' Mutual Confident Association, *Minutes*, Annual Delegate Meeting, 19 May 1884.
205. Wilson, *Autobiography*, 102.
206. Benney, *Charity Main*, 87.
207. Campbell, 'Slaves', 202.
208. B. Pickard to *Miners' Advocate*, 14 Mar. 1874.
209. Challinor, *Lancashire Miners*, 117.
210. Douglass, *Pit Talk*, 41—2.
211. Northumberland Miners' Mutual Confident Association, *Minutes*,

Executive Committee Meeting, 22 July 1904.

212. Northumberland Miners' Mutual Confident Association, *Minutes*, Delegate Meeting, 19 June 1875.

213. Northumberland Miners' Mutual Confident Association, *Minutes*, Annual Delegate Meeting, 15 Nov. 1884; Committee Meeting, 5 July 1886; Neville, 'Yorkshire Miners', 272; *Colliery Guardian*, 18 Feb. 1910; Hair, 'Social History', 178–9, 426; Duckham. *Scottish Industry*, 311.

214. *Colliery Guardian*, 18 June 1880.

215. *Colliery Guardian*, 24 Mar., 28 Apr. 1905.

216. Griffin, *Coalmining Industry*, 93–4; Burgess, *Industrial Relations*, 195, 199, 203, 209; Challinor, *Lancashire Miners*, 192; Rowe, *Wages*,. 39; Arnot, *South Wales*, 92; *Colliery Guardian*, 12 Jan., 31 Aug. 1900; Webbs, *Trade Unionism*, 323–4.

217. Jevons, *Coal Trade*, 338–9.

218. Williams, *Digging*, 28–30. See also National Union of Mineworkers (North Wales Area), Autobiography of E. Hughes, 1856–1925.

219. Jevons, *Coal Trade*, 348.

220. *Colliery Guardian*, 23 Mar. 1900.

221. *Colliery Guardian*, 20 Jan. 1905.

222. Williams, *Derbyshire Miners*, 219, 277.

223. Northumberland Miners' Mutual Confident Association, *Minutes*, Committee Meeting, 5 Feb. 1885; Williams, *Derbyshire Miners*, 277, 374.

224. Point of Ayr Lodge, Minutes, Committee Meeting, 25 Jan. 1894.

225. *Colliery Guardian*, 30 June 1905.

226. Benson, in *EHR* XXXI/ 3, 417.

227. Chaplin, *Durham Villages*, 21–2.

CONCLUSION
(pp. 214–215)

1. E. P. Thompson, *The Making of the English Working Class*, Harmondsworth 1968, 13.

2. V. Gilbert, 'Theses and Dissertations', *Bulletin of the Society for the Study of Labour History* 35 (Autumn 1977); V. Gilbert, 'Theses and Dissertations', *Bulletin of the Society for the Study of Labour History* 37 (Autumn 1978).

3. For the debate on the labour aristocracy see E. J. Hobsbawm, *Labouring Men: Studies in the History of Labour*, London 1964; H. Pelling, *Popular Politics and Society in Late Victorian Britain*, London 1968.

4. Such a conclusion calls into question some of the generalisations commonly made about the determinants of working-class thrift, about the pervasiveness of thrift among other occupational groups and, indeed, among the working-class as a whole.

Bibliography

PRIMARY AUTHORITIES: MANUSCRIPT

Bell, W., 'Reminiscences' (Durham Record Office).

Brown, Billy, Transcript of interview, 1976 (Northumberland Record Office).

City and County of Durham Permanent Benefit Building Society, Records (Durham Record Office).

Cowen, Ned, 'Of Mining Life And Aal Its Ways' (Durham Record Office).

Dalkin, T., 'Story of 50 Yrs, A Miner, And Soldier' (Durham Record Office).

Goodley, Harold, 'Bewicke Main. The Colliery Village Which Disappeared' (Durham Record Office).

Hodges, A., 'I remember' (Durham Record Office).

Hughes, Edward, 'Autobiography' (National Union of Mineworkers [North Wales] Areas, Wrexham).

Larmer, David, Ruby and Rose, Transcript of interview, 1976 (Durham Record Office).

McBurnie, George, 'Washington 1910–1974 (Durham Record Office).

McBurnie, George, 'Washington Old and New 1910–' (Durham Record Office).

Point of Ayr Lodge of the Denbighshire and Flintshire Miners' Federation, Minutes (National Union of Mineworkers [North Wales area], Wrexham).

Scottish Miners' Federation, Minutes (National Library of Scotland).

Stonelake, Edmund, 'autobiography' (University College of Swansea).

Sutherland Estate Records (Staffordshire Record Office).

West Lothian District, Minutes (National Library of Scotland).

INTERVIEWS

Interview with H. J. Evans, Essington, 1977.

Interview with A. J. R. Hickling, Sedgley, 1977.

PRIMARY AUTHORITIES: PRINTED

Central Association for Dealing with Distress Caused by Mining Accidents, *Annual Reports*.

Independent Order of Odd-Fellows, Manchester Unity Friendly Society. Supplementary Report, July 1st, 1872.

Lanarkshire Miners' County Union, *Minutes* (National Library of Scotland).

Lancashire and Cheshire Miners' Federation, *Minutes* (National Union of Mineworkers [North Western Area], Bolton).

Northumberland and Durham Miners' Permanent Relief Fund, *Annual Reports.*

Northumberland Miners' Mutual Confident Association, *Minutes* (National Union of Mineworkers [Northumberland Area], Newcastle).

Report of the Whickham Urban District Council as to the Condition of Marley Hill and District . . . (Durham Record Office).

South Wales Miners' Federation, *Minutes* (University College of Swansea).

Transactions and Results of the National Association of Coal, Lime and Iron-stone Miners . . ., 1864.

Wakefield Public Library, Newspaper Cuttings Collection.

West Riding Miners' Permanent Relief Fund, *Annual Reports.*

OFFICIAL PUBLICATIONS

Children's Employment Commission, 1842.

Reports of the Commissioner appointed under the Provisions of the Act 5 and 6 Vict. c.99 . . ., 1844—59.

Reports of H.M. Inspectors of Mines, 1851—1914.

Select Committee on Accidents in Coal Mines, 1852—3.

Select Committee on Acts for the Regulation and Inspection of Mines, 1866.

Board of Trade, Reports on Trade Unions, 1896—7.

Royal Commission on the Poor Laws and Relief of Distress, 1909—10.

NEWSPAPERS AND PERIODICALS

(a)   National
British Miner And General Newsman
Colliery Guardian
Labour Tribune
The Miner
Miners' Advocate
Miner and Workman's Advocate
Miner and Workmen's Examiner
Mining Journal
Post Magazine and Insurance Monitor
The Provident

(b)   Local
Accrington Times
Barnsley Chronicle
Barnsley Independent

Bolton Journal
Dean Forest Mercury
Derbyshire Courier
Derbyshire Times
Glamorgan Free Press
Hamilton Advertiser
Liverpool Mercury
Mexboro' & Swinton Times
Newcastle Weekly Chronicle
North Wales Miners' Magazine
Nuneaton Chronicle
Sheffield Daily Telegraph
Sheffield Independent
Staffordshire Sentinel
Wakefield Express
Western Mail
Wigan Observer
Wolverhampton Chronicle
Yorkshire Post

AUTOBIOGRAPHIES

Bullock, Jim, *Bowers Row: Recollections of a mining village*, Wakefield 1976.

Chaplin, Sid, *Durham Mining Villages*, Durham 1971.

Coombes, B. L., *These Poor Hands: The Autobiography of a Miner Working in South Wales*, London 1939.

Durland, Kellogg, *Among the Fife Miners*, London 1904.

Halliday, Joseph, *Just Ordinary, But . . . an Autobiography*, Waltham Abbey 1959.

Holman, Bob, *Behind the Diamond Panes: The Story of a Fife Mining Community*, Cowdenbeath 1952.

Horner, Arthur, *Incorrigible Rebel*, London 1960.

Lawson, Jack, *Who Goes Home?*, London 1945.

Lee, Jennie, *This Great Journey: A Volume of Autobiography 1904–45*, London 1963.

Parfitt, A. J. *My Life as a Somerset Miner*, Radstock 1930.

Parkinson, George, *True Stories of Durham Pit-Life*, London 1912.

Paynter, Will, *My Generation*, London 1972.

Redmayne, R. A.S., *Men, Mines, and Memories*, London 1942.

Ridyard, R., *Mining Days in Abram*, Leigh 1972.

Roberts, Robert, *The Classic Slum: Salford Life in the First Quarter of the Century*, Manchester 1971.

Rymer, E. A., 'The Martyrdom of the Mine, or, A 60 Years' Struggle for Life', ed R. G. Neville, *History Workshop* 1 (Spring 1976) and 2 (Autumn 1976).

Williams, T., *Digging for Britain*, London 1965.

Wilson, John, *Autobiography of Ald. John Wilson, J. P., M.P. A Record of a Strenuous Life*, Durham 1909.

OTHER AUTHORITIES LARGELY OR PARTLY PRIMARY
IN NATURE

The Colliery Year Book and Coal Trades Directory, London 1938.
Greenwell, G. C., *A Glossary of Terms Used in the Coal Trade of Northumberland and Durham*, London 1888.
Potts' Mining Register and Directory . . ., North Shields 1889 and 1910.
Wilson, T., *The Pitman's Pay and other Poems*, Gateshead 1843.

SECONDARY AUTHORITIES: BOOKS AND PAMPHLETS

Allen, G. C., *The Industrial Development of Birmingham and the Black Country 1860–1927*, London 1966.
Anderson, M., *Family Structure in Nineteenth Century Lancashire*, Cambridge 1971.
Arnot, R. Page, *A History of the Scottish Miners from the Earliest Times*, London 1955.
Arnot R. Page, *The Miners: A History of the Miners' Federation of Great Britain 1889–1910*, London 1949.
Arnot, R. Page, *South Wales Miners Glowyr de Cymru: A History of the South Wales Miners' Federation (1898–1914)*, London 1967.
Ashton, T. S. and Sykes, J., *The Coal Industry of the Eighteenth Century*, Manchester 1964.
Benney, Mark, *Charity Main: A Coalfield Chronicle*, London 1946.
Benson, John and Neville, R. G., eds, *Studies in the Yorkshire Coal Industry*, Manchester 1976.
Buchanan, R. A. and Cossons, N., *The Industrial Archaeology of the Bristol Region*, Newton Abbot 1969.
Bulman, H. F., *Coal Mining and the Coal Miner*, London 1920.
Burgess, K., *The Origins of British Industrial Relations: The Nineteenth Century Experience*, London 1975.
Burton, Anthony, *The Miners*, London 1976.
Campbell, G. L., *Miners' Insurance Funds: Their Origin and Extent*, London 1880.
Challinor, Raymond, *Alexander MacDonald and the Miners*, London 1968.
Challinor, Raymond, *The Lancashire and Cheshire Miners*, Newcastle 1972.
Challinor, Raymond and B. Ripley, *The Miners' Association: A Trade Union in the age of the Chartists*, London 1968.
Checkland, S. G., *The Rise of Industrial Society in England*, London 1964.
Clapham J. H., *An Economic History of Modern Britain: Free Trade and Steel 1850–1886*, Cambridge 1932.
Clegg, H. A., Fox, A. and Thompson, A. F., *A History of British Trade Unions since 1889*, I, Oxford 1964.
Coates, B. E., and Lewis, G. M., *The Doncaster area*, Sheffield 1966.
Cohen, J. L., *Workmen's Compensation in Great Britain*, London 1923.

Colls, Robert, *The Collier's Rant: Song and Culture in the Industrial Village*, London 1977.

Corfe, T. H., Davies, R. W. and Walton, A., *History Field Studies in the Durham Area: A Handbook*, Durham 1967.

Dallas, Karl, *One Hundred Songs of Toil*, London 1974.

Daunton, M. J., *Coal Metropolis: Cardiff 1870–1914*, Leicester 1977.

Davison, Jack, *Northumberland Miners 1919–1939*, Newcastle 1973.

Dennis, N., Henriques, F. and Slaughter, C., *Coal is our Life: An Analysis of a Yorkshire Mining Community*, London 1956.

Douglass, Dave, *Pit Talk in County Durham: A Glossary* . . , Oxford 1973.

Down, C. G. and Warrington, A. J., *The History of the Somerset Coalfield*, Newton Abbot 1971.

Duckham, B. F., *A History of the Scottish Coal Industry, I, 1700–1815: A Social and Industrial History*, Newton Abbot 1970.

Duckham, Helen and Baron, *Great Pit Disasters: Great Britain 1700 to the Present Day*, Newton Abbot 1973.

Elsas, M., ed., *Iron in the Making: Dowlais Iron Company Letters 1782–1860*, Cardiff 1960.

Evans, Chris, *Blaencwmdulais: A Short History of the Social & Industrial Development of Onllwyn and Banwen-Pyrddin*, Cardiff 1977.

Evans, Chris, *The Industrial and Social History of Seven Sisters*, Cardiff 1964.

Foster, John, *Class Struggle and the Industrial Revolution: Early Industrial Capitalism in three English towns*, London 1974.

Fynes, Richard, *The Miners of Northumberland and Durham*, Sunderland 1873.

Galloway, Robert L , *Annals of Coal Mining and the Coal Trade* 2 vols, Newton Abbot 1971.

Galloway, Robert L., *A History of Coal Mining in Great Britain*, Newton Abbot 1969.

Gosden, P. H. J. H., *Self-Help: Voluntary Associations in Nineteenth-Century Britain*, London 1973.

Gregory, Roy, *The Miners and British Politics 1906–1914*, Oxford 1968.

Griffin, A. R., *The British Coalmining Industry: Retrospect and Prospect*, Buxton 1977.

Griffin, A. R., *Mining in the East Midlands 1550–1947*, London 1971.

Griffiths, J. H. and Rundle, Frank, *The Pit on the Downs: A History of Eppleton Colliery 1825–1975*, Houghton-le-Spring n.d.

Hamilton, H., *The Industrial Revolution in Scotland*, London 1966.

Hammond, J. L. and B., *The Skilled Labourer 1760–1832*, London 1919.

Harris, José, *William Beveridge: A Biography*, Oxford 1977.

Harrison, Brian, *Drink and the Victorians: The Temperance Question in England 1815–1872*, London 1971.

Hetton Town Council, *Hetton Town: Official Guide*, Hetton-le-Hole n.d.

Hobsbawn, E. J., *Labouring Men: Studies in the History of Labour*, London 1964.

Horn, Pamela, *The Rise and Fall of the Victorian Servant*, Dublin 1975.

Horne, H. O., *A History of Savings Banks*, Oxford 1947.

Jevons, H. S., *The British Coal Trade*, London 1915.

John, A. H., *The Industrial Development of South Wales 1750–1850*, Cardiff 1950.

Jones, P. N., *Colliery Settlement in the South Wales Coalfield 1850 to 1926*, Hull 1969.

Kirby, M. W., *The British Coalmining Industry 1870–1946: A Political and Economic History*, London 1977.

Lawrence, D. H., *The Complete Short Stories* I, London 1955.

Lewis, Brian, *Coal Mining in the Eighteenth and Nineteenth Centuries*, London 1971.

Lewis, R. A., *Coal Mining in Staffordshire*, Stafford 1966.

Llewllyn, T. L. *Miners' Nystagmus: Its Causes and Prevention*, London 1912.

Lloyd, A. L., *Come all ye Bold Miners: Ballads and Songs of the Coalfields*, London 1952.

Lloyd, A. L., *Folk Song in England*, St Albans 1975.

Lloyd, Edward, *History of the Crook and Neighbourhood Co-operative Corn Mill, Flour & Provision Society Limited*, Pelaw-on-Tyne 1916.

Longmate, Norman, *The Waterdrinkers: A History of Temperance*, London 1968.

Machin, Frank, *The Yorkshire Miners: A History* I, Barnsley 1958.

Malcomson, R. W., *Popular Recreations in English Society 1700–1850*, Cambridge 1973.

Mathias, Peter, *The First Industrial Nation: An Economic History of Britain 1700–1914*, London 1969.

Mee, Graham, *Aristocratic Enterprise: The Fitzwilliam Industrial Undertakings 1795–1857*, Glasgow 1975.

Millman, R. N., *The Making of the Scottish Landscape*, London 1975.

Mitchell, B. R. and Deane, P., *Abstract of British Historical Statistics*, Cambridge 1962.

Moore, Robert, *Pit-Men, Preachers & Politics: The Effects of Methodism in a Durham Mining Community*, Cambridge 1974.

Morgan, J. E., *A Village Workers Council and what it Accomplished: Being a Short History of the Lady Windsor Lodge, South Wales Miners' Federation*, Pontypridd 1956.

Morgan, K. O., *Keir Hardie: Radical and Socialist*, London 1975.

Morrah, D., *A History of Industrial Life Assurance*, London 1955.

Morris, J. H. and Williams, L. J., *The South Wales Coal Industry 1841–1875*, Cardiff 1958.

M'Vail, John C., *Housing of Scottish Miners: Report on the Housing of Miners in Stirlingshire and Dunbartonshire*, Glasgow 1911.

National Museum of Wales, *Welsh Coal Mines*, Cardiff 1976.

Pelling, Henry, *Popular Politics and Society in Late Victorian Britain*, London 1968.

Perry, P. J., *A Geography of 19th-Century Britain*, London 1975.

Phelps Brown, E. H., *The Growth of British Industrial Relations: A*

Study from the Standpoint of 1906–14, London 1959.

Pollard, Sidney, *The Genesis of Modern Management: A Study of the Industrial Revolution in Great Britain*, Harmondsworth 1968.

Potts, A. and Wade, E., *Stand True*, Newcastle n.d.

Raybould, T. J., *The Economic Emergence of the Black Country: A Study of the Dudley Estate*, Newton Abbot 1973.

Raynes, J. R., *Coal and its Conflicts: A Brief Record of the Disputes between Capital & Labour in the Coal Mining Industry of Great Britain*, London 1928.

Rowe, J. W. F., *Wages in the Coal Industry*, London 1923.

Rubinstein, David, *Victorian Homes*, Newton Abbot 1974.

Russell, J. F. and Elliott, J. H., *The Brass Band Movement*, London 1936.

Samuel, Raphael, ed., *Miners, Quarrymen and Saltworkers*, London 1977.

Sayers, R. S., *A History of Economic Change in England, 1880–1939*, London 1967.

Shorter, Edward, *The Making of the Modern Family*, Glasgow 1977.

Smith, W., *An Economic Geography of Great Britain*, London 1949.

Smith, W., *An Historical Introduction to the Economic Geography of Great Britain*, London 1968.

Sturgess, R. W., *Aristocrat in Business: The Third Marquis of Londonderry as Coalowner and Portbuilder*, Durham 1975.

Supple, B. E., *The Royal Exchange Assurance: A History of Insurance, 1720–1970*, Cambridge 1970.

Taylor, A. J., ed., *The Standard of Living in Britain in the Industrial Revolution*, London 1975.

Trinder, Barrie, *The Industrial Revolution in Shropshire*, Chichester 1973.

Tunstall, Jeremy, *The Fishermen*, London 1962.

Tylecote, M., *The Mechanics' Institutes of Lancashire and Yorkshire before 1851*, Manchester 1957.

Walvin, James, *Leisure and Society 1830–1950*, London 1978.

Webb, Sidney, *The Story of the Durham Miners (1662–1921)*, London 1921.

Webb, Sidney and Beatrice, *The History of Trade Unionism*, London 1894.

White, J. W. and Simpson, R., *Jubilee History of West Stanley Co-operative Society Limited, 1876 to 1926*, Pelaw-on-Tyne 1926.

Williams, J. E., *The Derbyshire Miners: A Study in Industrial and Social History*, London 1962.

Wilson, A. and Levy, H., *Workmen's Compensation* 2 vols, Oxford 1939.

Zeldin, Theodore, *France 1848–1945*, 1, *Ambition, Love and Politics*, Oxford 1973.

ARTICLES

Ball, F. J., 'Housing in an Industrial Colony: Ebbw Vale, 1778–1914', *The History of Working-Class Housing: A Symposium*, ed. S. D. Chapman, Newton Abbot 1971.

Benson, John, 'Colliery Disaster Funds, 1860—1897', *International Review of Social History* XIX/1 (1974).

Benson, John, 'English Coal-Miners' Trade-Union Accident Funds, 1850—1900', *Economic History Review* XXVIII/3 (Aug. 1975).

Benson, John, 'The Motives of 19th-Century English Colliery Owners in Promoting Day Schools', *Journal of Educational Administration and History* III/I (Dec. 1970).

Benson, John, 'Non-fatal coalmining accidents', *Bulletin of the Society for the Study of Labour History* 32 (Spring 1976).

Benson, John, 'The Thrift of English Coal-Miners, 1860—95', *Economic History Review* XXXI/3 (Aug. 1978).

Bulmer, M.I.A., 'Sociological Models of the Mining Community', *Sociological Review* 23/1 (Feb. 1975).

Colls, Robert, ' "Oh Happy English Children" Coal, Class and Education in the North-East', *Past and Present* 73 (Nov. 1976).

Delves, T., 'Popular Recreations and their Enemies', *Bulletin of the Society for the Study of Labour History* 32 (Spring 1976).

Goffee, R. E., 'The Butty System and the Kent Coalfield', *Bulletin of the Society for the Study of Labour History* 34 (Spring 1977).

Francis, Hywel, 'The Origins of the South Wales Miners' Library', *History Workshop* 2 (Autumn 1976).

Gilbert, V., 'Theses and Dissertations', *Bulletin of the Society for the Study of Labour History* 35 (Autumn 1977) and 37 (Autumn 1978).

Haines, M. R., 'Fertility, Nuptiality, and Occupation: A Study of Coal Mining Populations and Regions in England and Wales in the Mid-Nineteenth Century', *Journal of Interdisciplinary History* VIII/2 (Autumn 1977).

Hair, P. E. H., 'Mortality from Violence in British Coal-Mines, 1800—50', *Economic History Review* XXI/3 (1968).

Hilton, G. W., 'The Truck Act of 1831', *Economic History Review* X/3 (1958).

John, A. V., 'The Lancashire Pit Brow Lasses and the Campaign to remove Women from Surface Labour 1842—87', *Bulletin of the North West Group for the Study of Labour History* 3 (1976).

Jones, P. N., 'Workmen's Trains in the South Wales Coalfield 1870—1926', *Transport History* 3/1 (Mar. 1970).

Lambert, W. R., 'Drink and Work-Discipline in Industrial South Wales, c. 1800—1870', *Welsh History Review* VII/3 (June 1975).

Lambert, W. R., 'The Welsh Sunday Closing Act, 1881', *Welsh History Review* VI/2 (Dec. 1972).

Lawson, Jack, 'The Influence of Peter Lee', *Mining and Social Change: Durham County in the Twentieth Century*, ed. M. Bulmer, London 1978.

Lebon, J. H. G., 'The Development of the Ayrshire Coalfield', *Scottish Geographical Magazine* 49 (1933).

McCord, Norman and Rowe, D. J., 'Industrialisation and Urban Growth in North-East England', *International Review of Social History* XXII/1 (1977).

McCormick, B. and Williams, J. E., 'The Miners and the Eight-Hour Day, 1863–1913', *Economic History Review* XII/2 (1959).

McKenna, Frank, 'Victorian Railway Workers', *History Workshop* 1 (Spring 1976).

Meiklejohn, A., 'History of Lung Diseases of Coal Miners in Great Britain: Part II, 1875–1920', *British Journal of Industrial Medicine* 9/2 (1952).

Morris, J. H. and Williams, L. J., 'The Discharge Note in the South Wales Coal Industry, 1841–1898', *Economic History Review* X/2 (1957).

Mott, J., 'Miners, Weavers and Pigeon Racing', *Leisure and Society in Britain*, eds M. A. Smith, S. Parker and C. S. Smith, London 1973.

Neville, R. G. and Benson, John, 'Labour in the Coalfields (II): A Select Critical Bibliography', *Bulletin of the Society for the Study of Labour History* 31 (Autumn 1975).

Oren, L., 'The Welfare of Women in Labouring Families: England, 1860–1950', *Clio's Consciousness Raised: New Perspectives on the History of Women*, eds M. S. Hartman and L. Banner, New York 1974.

Outhwaite, R. B., 'Age at Marriage in England from the late Seventeenth to the Nineteenth Century', *Transactions of the Royal Historical Society* XXIII (1973).

Redmayne, R. A. S., 'Coal Mining', *Victoria County History of Worcester* 2, eds J. W. Willis-Bund and W. Page, London 1971.

Reid, D. A., 'The Decline of Saint Monday 1766–1876', *Past and Present* 71 (May 1976).

Rose, M. E., 'The Allowance System under the New Poor Law', *Economic History Review* XIX/3 (1966).

Rowley, John, 'Drink and the Public House in Urban Nottingham, 1830–1860', *Bulletin of the Society for the Study of Labour History* 32 (Spring 1976).

Storm-Clark, C., 'The Miners 1870–1970: A Test-Case for Oral History', *Victorian Studies* XV/1 (Sep. 1971).

Sturgess, R. W., 'Landownership, Mining and Urban Development in Nineteenth-Century Staffordshire', *Land and Industry: The Landed Estate and the Industrial Revolution*, eds J. T. Ward and R. G. Wilson, Newton Abbot 1971.

Taylor, A. J., 'Coal', *Victoria County History of Staffordshire* 2, eds M. W. Greenslade and J. G. Jenkins, London 1967.

Taylor, A. J., 'Labour Productivity and Technological Innovation in the British Coal Industry, 1850–1914', *Economic History Review* XIV/1 (1961).

Thomas, B., 'The Migration of Labour into the Glamorganshire Coalfield, 1861–1911', *Industrial South Wales 1750–1914: Essays in Welsh Economic History*, ed. W. E. Minchinton, London 1969.

Thorpe, E., 'Politics and Housing in a Durham Mining Town', *Mining and Social Change: Durham County in the Twentieth Century*, ed. M. Bulmer, London 1978.

Trist, E. L. and Bamforth, K. W., 'Some Social and Psychological Consequences of the Longwall Method of Coal-getting', *Human Relations* 4/1 (1951).

Walters, Rhodri, 'Labour Productivity in the South Wales Steam-Coal Industry 1870–1914', *Economic History Review* XXVIII/2 (May 1975).

Williams, J. E., 'Labour in the Coalfields: A Critical Bibliography', *Bulletin of the Society for the Study of Labour History* 4 (Spring 1962).

Worpell, J. G., 'The Iron and Coal Trades of the South Staffordshire Area Covering the Period 1845–1853', *West Midland Studies* 9 (1976).

UNPUBLISHED THESES

Allan, Kenneth, 'The Recreations and Amusements of the Industrial Working Class, in the Second Quarter of the Nineteenth Century with Special Reference to Lancashire', (Manchester University M.A. thesis, 1947).

Barnsby, G. J., 'Social Conditions in the Black Country in the Nineteenth Century', (Birmingham University Ph.D. thesis, 1969).

Benson, John, 'The Compensation of English Coal Miners and their Dependants for Industrial Accidents, 1860–97', (Leeds University Ph.D. thesis, 1974).

Campbell, Alan B., 'Honourable Men and Degraded Slaves: A Social History of Trade Unionism in the Lanarkshire Coalfield, 1775–1874, with Particular Reference to the Coatbridge and Larkhall Districts', (Warwick University Ph.D. thesis, 1976).

Evison, J., 'Conditions of Labour in Yorkshire Coal-Mines, 1870–1914', (Birmingham University B.A. thesis, 1963).

Evison, J., 'The Opening up of the "Central" Region of the South Yorkshire Coalfield and the Development of its Townships as Colliery Communities 1875–1905', (Leeds University M.Phil. thesis, 1972).

Hair, P. E. H., 'The Social History of British Coalminers 1800–1845', (Oxford University D.Phil. thesis, 1955).

Hassan, J. A., 'The Development of the Coal Industry in Mid and West Lothian 1815–1873', (Strathclyde University Ph.D. thesis, 1976).

John, A. V., 'Women Workers in British Coal Mining, 1840–90, with Special Reference to West Lancashire', (Manchester University Ph.D. thesis, 1976).

Jones, M., 'Changes in Industry Population and Settlement on the Exposed Coalfield of South Yorkshire 1840–1908', (Nottingham University M.A. thesis, 1966).

Jones, P. N., 'Aspects of the Population and Settlement Geography of the South Wales Coalfield 1850–1926', (Birmingham University Ph.D. thesis, 1965).

Neville, R. G., 'The Yorkshire Miners, 1881–1926: A Study in Labour and Social History', (Leeds University Ph.D. thesis, 1974).

Seeley, J. Y. E., 'Coal Mining Villages of Northumberland and Durham:

A Study of Sanitary Conditions and Social Facilities, 1870—1880', (Newcastle University M.A.thesis, 1973).

Taylor, Eric, 'The Working Class Movement in the Black Country 1863—1914', (Keele University Ph.D. thesis, 1974).

Turner, Margaret, 'The Miner's Search for Self-Improvement: The History of Evening Classes in the Rhondda Valley from 1862—1914', (Wales University M.A. thesis, 1966).

Wilson, G. M., 'The Miners of the West of Scotland and their Trade Unions, 1842—74', (Glasgow University Ph.D. thesis, 1977).

Youngson Brown, A. J., 'The Scots Coal Industry, 1854—1886', (Aberdeen University D.Litt. thesis, 1952).

Index

273